QUANTUM MECHANICS

QUANTUM MECHANICS
Axiomatic Theory with Modern Applications

Nelson Bolivar, PhD

Gabriel Abellán, BSc

Apple Academic Press Inc.
3333 Mistwell Crescent
Oakville, ON L6L 0A2 Canada

Apple Academic Press Inc.
9 Spinnaker Way
Waretown, NJ 08758 USA

© 2018 by Apple Academic Press, Inc.

First issued in paperback 2021

No claim to original U.S. Government works

ISBN 13: 978-1-77463-173-7 (pbk)
ISBN 13: 978-1-77188-691-8 (hbk)

Library and Archives Canada Cataloguing in Publication

Bolivar, Nelson E., author
Quantum mechanics: axiomatic theory with modern applications / Nelson E. Bolivar, PhD, Gabriel Abellán, BSc.
Includes bibliographical references and index.
Issued in print and electronic formats.
ISBN 978-1-77188-691-8 (hardcover).--ISBN 978-1-351-16628-7 (PDF)
1. Quantum theory. I. Abellán, Gabriel (Physics professor), author II. Title.
QC174.12.B65 2018   530.12   C2018-901448-2   C2018-901449-0

Library of Congress Cataloging-in-Publication Data

Names: Bolivar, Nelson E., 1978- author. | Abellán, Gabriel (Physics professor), author.
Title: Quantum mechanics: axiomatic theory with modern applications / Nelson E. Bolivar, PhD, Gabriel Abellán, BSc.
Description: Oakville, ON; Waretown, NJ : Apple Academic Press, [2018] |
Includes bibliographical references and index.
Identifiers: LCCN 2018008420 (print) | LCCN 2018012752 (ebook) | ISBN 9781351166287 (ebook) | ISBN 9781771886918 | ISBN 9781771886918 (hardcover ; alk. paper) | ISBN 1771886919 (hardcover ; qalk. paper) | ISBN 9781351166287 (eBook) | ISBN 135116628X (eBook)
Subjects: LCSH: Quantum theory. | Quantum theory--Mathematics. | Axiomatic set theory.
Classification: LCC QC174.12 (ebook) | LCC QC174.12 .B65 2018 (print) | DDC 530.12--dc23
LC record available at https://lccn.loc.gov/2018008420

Apple Academic Press also publishes its books in a variety of electronic formats. Some content that appears in print may not be available in electronic format. For information about Apple Academic Press products, visit our website at **www.appleacademicpress.com** and the CRC Press website at **www.crc-press.com**

*To those who were here,
are here, and will be here
with us.*

ABOUT THE AUTHORS

Nelson Bolivar, PhD

Professor in Physics, Central University of Venezuela, Faculty of Science, School of Physics, Venezuela, E-mail: nelson.bolivar@ciens.ucv.ve

Nelson Bolivar is currently a Physics Professor in the Physics Department at the Universidad Central de Venezuela, where he has been teaching since 2007. His interests include quantum field theory applied in condensed matter and AdS/CMT correspondence. He obtained his PhD in physics from the Université de Lorraine (France) in 2014 in a joint PhD with the Universidad Central de Venezuela. His BSc in physics is from the Universidad Central de Venezuela.

Gabriel Abellán, BSc

Professor in Physics, Central University of Venezuela, Faculty of Science, School of Physics, Venezuela, E-mail: gabriel.abellan@ciens.ucv.ve

Gabriel Abellán is a Professor of Physics in the Physics Department at the Universidad Central de Venezuela. He has taught several courses since 2013, including Classical Mechanics, Waves and Optics, and Statistical Physics, among others. He received his BS degree at the Universidad Central de Venezuela and is currently doing his doctorate research under Professor Nelson Bolivar's supervision. Professor Abellán has collaborated with Santillana Publishers (Venezuela) in the writing, revision, and correction of several physics books used in high school education. His current interests are thermal field theories, gauge-gravity duality, and relationships between the theory of dynamical systems and quantum field theory, particularly regarding some aspects of renormalization. Professor Gabriel Abellán is an active science communicator and writes an informal blog about general science curiosities. In addition, he also conducts the choir of the Science Faculty at the Universidad Central de Venezuela that performs regularly in Caracas.

CONTENTS

LIST OF ABBREVIATIONS

ABC	antiperiodic BCs
AO	atomic orbitals
BCs	boundary conditions
BJTs	bipolar junction transistors
CdSe	cadmium selenide
CPP	current perpendicular to the plane
CPP	perpendicular plane current
DMS	diluted magnetic semiconductors
DOS	density of states
EMA	effective mass approximation model
FETs	field effect transistors
FM	ferromagnet
GaAs	gallium arsenide
GaAsAl	Arsenide de Galium-Aluminum
GMR	giant magneto resistance
HOMO	highest occupied molecular orbital levels
IEC	interlayer exchange coupling
IQHE	integer quantum hall effect
ISO	intrinsic spin orbit
LCAO	linear combination of atomic orbit theory
LEDs	light emitting devices

LUMO	lowest unoccupied molecular orbital levels
MBE	molecular beam epitaxy
MO	molecular orbital
MR	magnetoresistance
MTJs	magnetic tunnel junctions
NIR	near infrared spectral region
PbS	galena
PGP	pretty good privacy
PL	photoluminescence
QDots	quantum dots
RKKY	Raudermann, Kittel, Kasuya, and Yoshida
RSO	Rashba spin orbit
spin-LEDs	spin-dependent LEDs
SSL	secure socket layer
TMR	tunneling magnetoresistance
TRS	time reversal symmetry
UC	U-CONTROLLED
VPN	virtual private networking

LIST OF SYMBOLS

k_B	Boltzmann constant
$\oint$	closed integral
$\downarrow$	denotes spin down
$\uparrow$	denotes spin up
$\dot{a}(\vec{r}, t)$	derivative with respect to a parameter t
$a'(\vec{r})$	derivative with respect to space
$\otimes$	direct product
$\oplus$	direct sum
m_e	electron mass
$\langle \psi | \phi \rangle$	inner product in Hilbert space
Δ_{SO}	intrinsic spin-orbit strength
$\langle A \rangle$	mean value of the object A
$\hat{A}, \mathbf{A}$	operators or linear maps
$\parallel$	parallel to (...)
$\perp$	perpendicular to (...)
I_c	persistent charge currents
J_c	persistent charge density currents
I_s	persistent spin currents
J_s	persistent spin density currents
$\prod_{a=c}^{b}$	productory
$| \psi \rangle \langle \phi |$	projector on Hilbert space

m_p	proton mass
$\psi(t, \vec{r})$	quantum state in the position representation
λ_R	Rashba spin-orbit coupling strength
$\hbar$	reduced Planck's constant
c	speed of light
$\sum_{a=c}^{c}$	sumatory or series
t, τ	time parameters
$\langle \psi |$	vector on dual space
$| \psi \rangle$	vector on Hilbert space
$\vec{A}$	vector quantity or vector field

LIST OF FIGURES

PREFACE

According to the new conception of mechanics, the radiation, previously characterized by its continuity, was reduced to material granules or discrete amounts of energy. However, in defining steady states of the electron, a simultaneous wave character was attributed to it, the momentum of the electron had to be matched by a wavelength, so that Planck's constant, which had served to introduce the electron corpuscular character in the theory of radiation, allowed also to transfer the wave nature to the material corpuscles.

The traditional conception of the electron, which considered it as a simple point charge in a medium without structure, was discarded and it was necessary to accept, on the contrary, that the electron in movement is always accompanied by a series of waves that, determine the direction to follow.

The German Werner Heisenberg solved the problem of determining the nature of the wave associated with the electron with a probabilistic interpretation, according to the so-called uncertainty principle. According to this result, the product of the uncertainties or inaccuracies with which two associated quantities are known, that is, pairs of magnitudes in which it happens that the better one is to measure a more imprecise one, the other is to be of the order of the magnitude of the of Planck's constant.

The essence of this uncertainty principle is understood by considering that, when performing a measurement on a particle, it is impossible not to keep un-change the state of the particle. For example, if the electron is to be visualized to order to study it, the light used would radically change its physical state.

In particular cases, but it is not possible to determine its exact position simultaneously. Thus, a probability distribution can only be given for the various possible situations. By applying quantum mechanics to the study of the atom the deterministic orbits of the first atomic models disappear and are replaced by the expressions of probability or wave functions devised by Erwin Schrödinger.

Developed with these guidelines, quantum mechanics not only eliminated the large logical difficulties presented by theoretical classical physics, but also allowed to solve new problems, such as the interpretation of valence forces and intermolecular forces. This book aims to ride the reader through the various aspects of quantum mechanics. It no only involves the basics, but along the years its been a fruitful subject to include so many topics. These goes from electronics, spintronics, cryptography and other more theoretical aspects as path integral formulation and supersymmetric quantum mechanics.

—Nelson Bolivar

ACKNOWLEDGMENTS

We would like to thank Dr. J. A. Lopez for his help in preparing this manuscript, and Arley Larrota for the help with the graphics. Also, we would like to thank our colleagues for the frequent discussions and enthusiasms; and there is not a better support than family and friends.

—**Nelson Bolivar and Gabriel Abellán**

INTRODUCTION

It is well known that the purpose of classical mechanics is to determine the motion of a given physical system, reducing the study of a finite, but large, number of parameters and determining their behavior according to time. This problem is approached with by Newton's laws and the initial position and speed of the system. This determines the state of the system at each subsequent moment.

This classical description of every physical system has been shown poorly adapted to the physics in the end of the 18th century. Certain phenomena concerning size systems of the order of 10^{-6} meters could not be explained classically. The motion in classical systems is basically surrendered to a famous equations,

$$\vec{F} = m\vec{a}. \tag{I.1}$$

The second Newton's law. In the face of these "recalcitrant" phenomena a new set of equations was introduced, Schrödinger equations. This new mechanics was called "Quantum Mechanics" and succeeded to explain these phenomena, as well as being in agreement where classical mechanics was correct.

Quantum mechanics was basically developed in two periods. The first period begins with the introduction of the concept of action in 1900 due to Planck. In this period, the new mechanics was basically a mixture of classical concepts and non-classical ones, and was not considered completely satisfactory.

The second period, which began around 1925, is largely devoted to the result of progress of Heisenberg and Schrödinger. During this period the difficulties of the previous version of quantum mechanics are now completely resolved and this version receives the name of quantum mechanics.

The approach Schröderinger is based on the study of his equation, all dynamics comes from its evolution. Heisenberg's approach, also called matrix method, does have a slightly but related and equivalent approach, where Heisenberg equations are the rulers.

In fact, quantum mechanics, in its definitive version, describes the behavior of particles such as atoms, electrons, nuclei, molecules, photons, and so on. The description of sub-atomic particles requires further development of quantum mechanics what we called "quantum field theory" and begins essentially around 1947.

Matter is made up of molecules, which in turn are made of atoms. Atoms are never in equilibrium, they are composed of charged particles with different signs, and held together by Coulomb's law. Following classical mechanics, the electron, like a satellite orbiting by a planet, does not fall into the nucleus, but it is in motion along an orbit around the nucleus. On the other hand, Maxwell's theory states that when charged particles are accelerate they must emit electromagnetic radiation, so the electron around the nucleus should emit radiation and therefore lose energy and quickly collapse into the it, all these happening in around 10^{-10} seconds. It is evident that such a collapse would have terrible consequences, such as all chemistry could not work and life wouldn't exist. The fact that systems do not collapse after 10^{-10} seconds means that the electron orbiting does not emit radiation and therefore the classical explanation collapses.

Another interesting phenomena occurs when heating an element or is subject to a strong electric discharge. It happens that it emits light, an electromagnetic radiation. This radiation is formed only by a definite number of frequencies. Classical theory states that these frequencies must be all the same or a linear combination of the regular frequencies of periodic charge particles motions into atoms. More precisely, if ω_i are the elementary frequencies, then we expect to observe frequencies of the form $\omega = \sum_i n_i \omega_i$, where the n_i are positive integers. But, what is experimentally observed is that electromagnetic emission frequencies are of the form $\omega = \omega_n - \omega_m$ where ω_n is a fixed number of frequencies.

Consider electromagnetic radiation within an enclosed domain that is equilibrium with its environment. Historically, this radiation is also called

Figure I.1: Erwin Schrödinger. **Figure I.2:** Werner Heisenberg.

as black body radiation. If we think ideally, to make a small hole in the enclosure of a cavity, to allow no radiation to scape and be measured, then the radiation of the exterior has a chance to be absorb into the cavity, mimicking black body. The classical explanation leads to the conclusion that the number of electromagnetic waves with frequency between ω and $\omega + d\omega$ and give,

$$n(\omega)d\omega = \frac{8\pi\omega^2}{c^3}d\omega \qquad (I.2)$$

where c is the speed of light. In terms of the energy,

$$E(\omega)d\omega = \frac{8\pi\omega^2}{c^3}k_BTd\omega \qquad (I.3)$$

where T is the body temperature and k_B is the Boltzmann constant.

The latter in named Rayleigh-Jeans formula, and is in good agreement with the experiments for low frequencies. For high frequencies doesn't work, and it is not surprising, if we want to calculate the total energy we need to integrate the formula, and we found that it is a divergent integral.

This paradox was solved by Planck in 1900, by means of a radical proposal. He suggested that the radiation of a given frequency ω can only

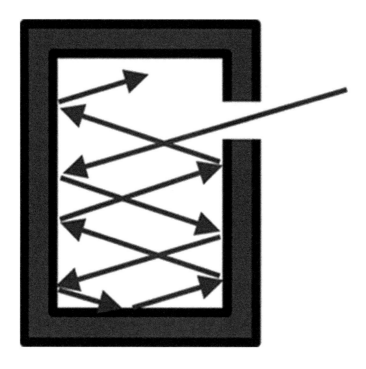

Figure I.3: A cavity that mimics a black body.

Figure I.4: Max Planck.   **Figure I.5:** Albert Einstein.

exchange energy with matter in discrete packets, each of energy $h\omega$, where h is a constant of fixed value, called Planck's constant and that has the units of an energy per time. Following this approach the law of distribution of the energy has the form,

$$E(\omega)d\omega = \frac{8\pi\omega^2}{c^3}\frac{1}{e^{\frac{h\omega}{k_B T}} - 1}d\omega \qquad (I.4)$$

known as Planck's law.

Note that for small ω or large T, we find Rayleigh-Jeans law. Planck's law correctly describes the frequency for all experiments and the total energy emitted. Since c is known and the total energy is experimentally measured, it is found that the value of the Planck constant is, $h = 6.55 \times 10^{-34}$ joules per second.

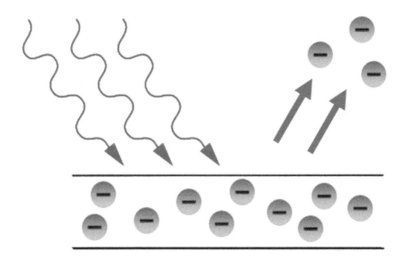

Figure I.6: Basic scheme of the photoelectric effect.

Another interesting and fundamental result occurs when an ultraviolet light strikes a metallic surface. It is observed experimentally an electron current, even when a stopping potential is present. If the retarding potential is large enough, then no electron emission is observed. The greatest potential for which electrons are observed is called V_s and is proportional to the maximum electron energy emitted by the radiated surface. Based on the classical formalism, electron energy is expected to increase with the intensity of ultra-violet light, and therefore, if light energy is ultraviolet, it should also increase the stopping potential. On the other hand, it has been experimentally found that stopping potential was independent of the intensity of light, but increased linearly with the frequency of light. This fact does not have a classical explanation, changing the frequency should not have any effect on the potential. The logical extension of Planck's quantum hypothesis was made in 1905 by Einstein to explain the photo-electric effect. He suggested not only that the energy exchange of energy and a material is carried out through packets, but that radiation energy only comes in discrete packets also, called photons, each of the energy $h\omega$. Based on this hypothesis the photoelectric effect can be explained; in fact, if the photon hitting the metal surface has higher energy than the work W needed to "tear apart" the electron of the atom, then the electron will emerge with energy $h\omega - W$ and then the stopping potential will be this difference. Then V_s will be linearly proportional to ω and the proportional constant is the constant h. It is remarkable to note that this proportionality constant can be experimental measured and coincides, in good agreement, with the constant found in the radiation of the black body.

Quantum mechanics has proven an enormously success, giving correct results in practically every situation to which it has been applied. However, an intriguing paradox still exist. In spite of the overwhelming practical success of quantum mechanics, the foundations contain unresolved problems, in particular, concerning the nature of measurement. An essential feature of quantum mechanics is that it is generally impossible, even in principle, to measure a system without disturbing it.

The detailed nature of this disturbance and the exact point at which it occurs are obscure and controversial. Thus, quantum mechanics attracted some of the best scientists of the 20th century, and working in the finest intellectual building of the period.

CHAPTER 1

INTRODUCTION TO SUPERPOSITION PRINCIPLE WAVES

1.1 WAVES SUPERPOSITION

The result of overlapping a couple of waveform is obtained by adding algebraically each of the sine waves that make up that complex motion. If we superpose sinusoidal waves of equal frequency, but with possible different amplitudes and phases, we will obtain another sine wave with the same frequency, but with different amplitude and phase. Eventually these waves can be canceled, for example, if they had equal amplitude but a phase difference of 180 degrees.

Of particular interest is the case of superposition of sine waves with different frequency and eventual different amplitude and phase.

Although the decomposition of any complex motion into an overlapping of different proportions of simple harmonic motions is strictly true for periodic complex motions, certain mathematical approximations allow us to decompose non-periodic motion into a set of simple motions as well.

The overlapping of sine waves whose frequencies have a simple integer relationship will result in a periodic complex motion. The next figures show the result of the superposition of different harmonics of a series.

Note that the resulting waveform depends on the amplitude and phase of each of the superimposed sine waves.

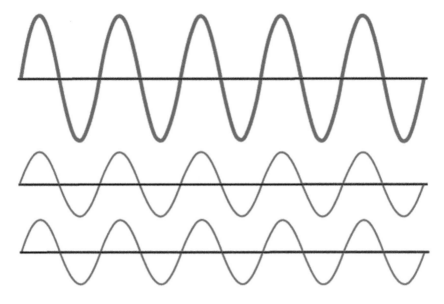

Figure 1.1: Results of the two superposition harmonic waves (the figure shows how two waves overlap to form a double amplitude wave (top)).

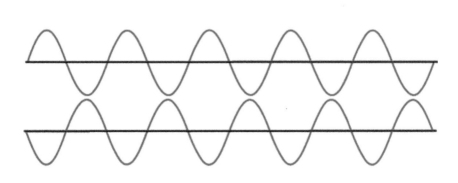

Figure 1.2: Results of the two superposition harmonic waves (the figure shows how a phase difference of 180 degrees cancels (top)).

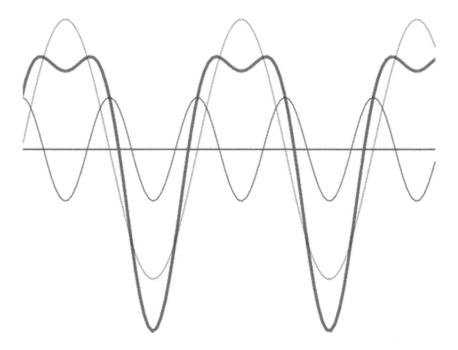

Figure 1.3: Result of the superposition two harmonic waves (red curve) with different phase, frequency, and amplitude.

1.1.1　Wave Equation

The sine that describes a waveform derives as a solution of a wave equation, a differential equation of second order in partial derivatives. One of the properties of the wave equation is that it is a linear equation, this means that it supports the principle of superposition, which means that if y_1 and y_2 are solutions of the same wave equation

$$\frac{\partial^2 y_1}{\partial x^2} - \frac{1}{v^2}\frac{\partial^2 y_1}{\partial t^2} = 0, \frac{\partial^2 y_2}{\partial x^2} - \frac{1}{v^2}\frac{\partial^2 y_2}{\partial t^2} = 0 \qquad (1.1)$$

that means nothing more than a couple of waves spread by the same string, then their sum is also solution

$$y = y_1 + y_2 \Rightarrow \frac{\partial^2 y}{\partial x^2} - \frac{1}{v^2}\frac{\partial^2 y}{\partial t^2} = 0 \qquad (1.2)$$

In the particular case of wave pulses, the result is that, although during the period of coincidence, deformation of the resulting wave can take strange forms, when both separate pulses continue without having been affected at all by the "collision" with each other.

In the case of sine waves, the principle is the same. However, since these waves range indefinitely in space, the coincidence occurs everywhere all the time. In addition, the shape of the resulting waves often possesses interpretation on itself.

1.1.1.1 *Same Sense, Frequency, and Amplitude*

We began assuming the simplest possible case: two waves propagated in a rope in the same direction, with same frequency and the same amplitude, differing only in their gap, that means they have a phase difference. We can write these two signals as

$$y_1 = A_0 \cos(\omega t - kx) \, y_2 = A_0 \cos(\omega t - kx + \phi), \qquad (1.3)$$

where have we taken the first sign as a reference and with offset 0. To talk about the gap between the two waves, rather than the angle ϕ is often used the distance Δx between a maximum of a wave and the maximum of the other, in such a way that refers to a outdated "half wavelength" or "a quarter wavelength," for example. The relationship between this gap and the angle ϕ is

$$\frac{\Delta x}{\lambda} = \frac{\phi}{2\pi} \qquad (1.4)$$

For these two signals, the overlap will be

$$y = A_0 \cos(\omega t - kx) + A_0 \cos(\omega t - kx + \phi), \qquad (1.5)$$

Applying the trigonometric relationship,

$$\cos(a) + \cos(b) = 2 \cos\left(\frac{a+b}{2}\right) \cos\left(\frac{a-b}{2}\right) \qquad (1.6)$$

the superposition of waves becomes,

$$y = 2A_0 \cos\left(\frac{\phi}{2}\right) \cos\left(\omega t - kx + \frac{\phi}{2}\right) = A(\phi) \cos\left(\omega t - kx + \phi'\right) \qquad (1.7)$$

with

$$A(\phi) = 2A_0 \cos\left(\frac{\phi}{2}\right) \qquad \phi' = \frac{\phi}{2} \qquad (1.8)$$

We interpret this result in a simple way: the sum of two waves traveling in the same direction with the same frequency and amplitude is a new wave traveling with a gap that is the average of the respective gaps and whose amplitude depends on that very same gap. The moment in which each wave starts it motion is absolutely relevant!

According to the value of the gap we have two important cases:

(i) *Constructive interference:* When the gap between waves is an even multiple of π, or in terms of the distance between peaks of the waves, if they are delayed in a integer number of wavelengths.

$$\phi = 2n\pi \Rightarrow \Delta x = n\lambda \qquad (1.9)$$

In this case, the waves are completely coincidental and the resulting wave is one in phase with them and with double amplitude,

$$A(0) = 2A_0 \qquad \phi'(0) = 0 \qquad (1.10)$$

It is said that these waves are in phase and have constructive interference.

Note that is twice the amplitude, the energy of the resulting waveform is the square of the component waves and so does power, that is, that the power of the sum is not equal to the sum of the energies.

(ii) *Destructive interference:* When the gap between waves is an odd multiple of π, or in terms of length, when they differ in a half integer multiple of the wavelength,

$$\phi = (2n + 1)\pi \quad \Rightarrow \Delta x = n\lambda \qquad (1.11)$$

If the waves are in this situation they cancel out each other and the result is a no wave, zero. Hence it's destructive interference.

$$A(\pi) = 0 \quad \phi' = \frac{\pi}{2} \qquad (1.12)$$

For this interference, the stored energy is naturally void and the same happens with the power. Again we see that the power of the sum is not the sum of the energies.

Other interference: when the waves are in an intermediate situation between being in phase or in phase opposition, the result is a wave whose amplitude is intermediate between 0 and $2A_0$. In particular, there is a lag for which the sum of the waves has exactly the same amplitude that each of the one added.

$$\phi = 0, \quad \phi = \pi, \quad \phi = \pi/2, \quad , \phi = 2\pi/3, \qquad (1.13)$$

Different Amplitude

Suppose now that we have two waves of the same frequency propagating in the same sense, but of different amplitude and phase:

$$y_1 = A_1 \cos(\omega t - kx), \quad y_2 = A_2 \cos(\omega t - kx + \phi), \qquad (1.14)$$

In this case, the use of trigonometric relations is not as simple as in the previous case, so first we will consider two simple cases and then give the general solution, with the help of the Phasor calculation. In physics and engineering, a phasor, named after the blend of phase and vector, is a complex number that represents a sinusoidal function whose amplitude, frequency, and phase are time-invariant.

Constructive interference: If the gap is zero, the two waves are in phase and the result is a wave with an amplitude that is the sum of both waves,

$$\phi = 0 \Rightarrow y = y_1 + y_2 = (A_1 + A_2)\cos(\omega t - kx), \qquad (1.15)$$

Destructive interference: If the waves are in opposition ($\phi = \pi$), there is a complete cancellation, but, for being one of the waves of greater amplitude than the other. The sum of both signals is,

$$\phi = \pi \Rightarrow y = y_1 + y_2 = (A_1 - A_2)\cos(\omega t - kx), \qquad (1.16)$$

If we take into account that the amplitudes are always positives, this overlap must be written correctly as

$$\phi = \pi \Rightarrow y = y_1 + y_2 = |A_1 - A_2|\cos(\omega t - kx + \phi'), \qquad (1.17)$$

with

$$\phi' = \begin{cases} 0 & A_1 > A_2 \\ \pi & A_1 < A_2 \end{cases}$$

In the general case of different amplitudes and arbitrary phase, we can make use of the Phasor calculation. Each of the signals can be put in the form

$$y_1 = A_1\cos(\omega t - kx) = \text{Re}\left(A_1 e^{j(\omega t - kx)}\right) = \text{Re}\left(\tilde{y}_1 e^{j(\omega t - kx)}\right)$$

$$y_2 = A_2\cos(\omega t - kx + \phi) = \text{Re}\left(A_2 e^{j(\omega t - kx + \phi)}\right) = \text{Re}\left(\tilde{y}_2 e^{j(\omega t - kx)}\right)$$

being each phasor

$$\tilde{y}_1 = A_1 \quad \tilde{y}_2 = A_2 e^{j\phi}, \tag{1.18}$$

When we add these two waves, the result will be a new wave, with a phasor form,

$$\tilde{y} = \tilde{y}_1 + \tilde{y}_2 = A_1 + A_2 e^{j\phi} \tag{1.19}$$

This sum of complex numbers is written as a single phasor that combines the amplitude and the gap of the resulting waveform,

$$\tilde{y} = A_1 + A_2 e^{j\phi} = A e^{j\phi'} \tag{1.20}$$

The amplitude of the oscillations will be the module of the complex number,

$$A = \sqrt{\tilde{y}\tilde{y}*} = \sqrt{(A_1 + A_2 e^{j\phi})(A_1 + A_2 e^{-j\phi})}$$
$$= \sqrt{A_1^2 + A_2^2 + A_1 A_2(e^{j\phi} + e^{-j\phi})} \tag{1.21}$$

and, by applying Euler's formula we get finally,

$$A = \sqrt{A_1^2 + A_2^2 + 2A_1 A_2\cos(\phi)} \tag{1.22}$$

This module is the vector sum of the two phasors, considered as vectors in the complex plane.

The two previous cases are contained in this result:

If $\phi = 0$, $\cos\phi = 1$, and the amplitude is reduced to

$$A = \sqrt{A_1^2 + A_2^2 + 2A_1 A_2} = \sqrt{(A_1 + A_2)^2} = A_1 + A_2 \tag{1.23}$$

if $\phi = \pi$, $\cos\phi = -1$, and is the amplitude

$$A = \sqrt{A_1^2 + A_2^2 - 2A_1 A_2} = \sqrt{(A_1 - A_2)^2} = |A_1 - A_2| \tag{1.24}$$

We can also obtain the previous case when both signals have the same amplitude, for any phase. If $A_1 = A_2 = A_0$

$$A = \sqrt{A_0^2 + A_0^2 + 2A_0^2 \cos(\phi)} = 2A_0 \sqrt{\frac{1 + \cos(\phi)}{2}} = 2A_0 \cos\left(\frac{\phi}{2}\right)$$
(1.25)

In addition to the amplitude, we get the offset of the resulting waveform as the argument of the phasor

$$\phi' = \arctan\left(\frac{\Im(\tilde{y})}{\Re(\tilde{y})}\right) = \arctan\left(\frac{A_2 \sin(\phi)}{A_1 + A_2 \cos(\phi)}\right)$$
(1.26)

In terms of the resulting wave energy, applying the formula for the energy contained in a wave we have,

$$E = \frac{\mu \omega^2 \lambda A^2}{2} = \frac{\mu \omega^2 \lambda}{2}(A_1^2 + A_2^2 + 2A_1 A_2 \cos(\phi))$$
$$= E_1 + E_2 + \mu \omega^2 \lambda A_1 A_2 \cos(\phi).$$
(1.27)

It reads that the power of the sum is equal to the sum of the energies, plus a term that depends on the amplitudes and the gap, this can be both positive, when there is constructive interference and negative when there is destructive interference.

1.1.1.2 Opposite Direction with Same Amplitude

We will now examine the case of two waves with same frequency and amplitude propagating in opposite directions:

$$y_1 = A \cos(\omega t - kx) \quad y_2 = A \cos(t + kx\omega),$$
(1.28)

In this case it is necessary to introduce the constant ϕ, because, for waves going in opposite directions the concept of gap doesn't make much sense. This constant can be included, but the results do not differ essentially from what you get without it.

The overlap of these two signals can be transformed by applying trigonometric relations

$$y = y_1 + y_2 = A \cos(\omega t - kx) + A \cos(\omega t + kx) = 2A \cos(\omega t) \cos(kx).$$
(1.29)

This is the equation in a stationary wave, which can be written in the form,

$$y = A(x)\cos(\omega t) \quad A(x) = 2A\cos(kx), \tag{1.30}$$

This equation tells us that although we have the overlap of two traveling waves, the sum is a wave in which all the points oscillate in phase with a position-dependent amplitude. Those points where is wider and negative, must be understood that the amplitude is the absolute value of this amplitude and the corresponding points are in phase opposition, that is, have a constant phase equal to π.

Nodes

The amplitude varies as a cosine, implying that there are points for which the amplitude of oscillation is null. These points are called nodes. The condition of zero amplitude gives the position of these nodes,

$$A(x_n) = 2A\cos(kx_n) = 0 \quad \Rightarrow kx_n = \frac{\pi}{2} + n\pi$$

$$\Rightarrow \Delta x = x_{n+1} - x_n = \frac{\pi}{k} = \frac{\lambda}{2}. \tag{1.31}$$

The distance between consecutive nodes is half wavelength.

Antinodes

The points in which the amplitude of oscillation is maximum are called antinodes. This bellies are the midpoints between nodes, and therefore the distance between consecutive bellies is also medium wavelength, and the distance of a belly to the next node is $\lambda/4$.

Different Amplitude

If the two waves do not have the same amplitude,

$$y_1 = A_1\cos(\omega t - kx) \quad y_2 = A_2\cos(\omega t + kx), \tag{1.32}$$

Suppose that $A_1 > A_2$. For the first wave we can write roughly,

$$y_1 = A_1\cos(\omega t - kx) = (A_1 - A_2)\cos(\omega t - kx) + A_2\cos(\omega t - kx), \tag{1.33}$$

With the second

$$y = y_1 + y_2 = (A_1 - A_2)\cos(\omega t - kx) + 2A_2\cos(\omega t)\cos(kx), \quad (1.34)$$

Comparing with that the two traveling waves we see they have the same sum than the latter.

CHAPTER 2

THE WAVE FUNCTION, EXPECTATIONS VALUES, AND UNCERTAINTY

2.1 THE FOUNDATIONS OF QUANTUM MECHANICS

2.1.1 Uncertainty Principle

The uncertainty principle, formulated by Heisenberg in 1927, is a fundamental principle in quantum mechanics. It forces us to review key concepts of classical physics, and from this analysis are born the concepts of quantum physics.

The principle states that in any physical system simultaneous measurement of two canonical conjugate variables q and p is subject to an intrinsic limitation of accuracy. If we denote with Δq and Δp inaccuracies of the two measures separately, they are related by the relation of indeterminacy,

$$\Delta p \Delta q \geq \hbar \tag{2.1}$$

To justify this on physical basis, several experimental ideas have been proposed, for example, the Heisenberg's microscope.

Imagine you want to measure the x coordinate and momentum p of an electron along the direction of motion. Suppose we know exactly the value of p before the measurement and want to determine x by observing the electron with a microscope, i.e., illuminating it with a beam of light that has a given direction and wavelength λ. Suppose a single photon being scatter by the electron at a point P and passing through the lens is focused at the point Q of a photographic film. The inaccuracy of measurement of x is at least equal to the resolving power of the microscope, which is given by,

$$\Delta x \approx \frac{\lambda}{\sin \alpha} \tag{2.2}$$

where α is the tilt of the cone subtended by the lens section at point P. On the other hand the photon going from P to Q can start from P in any direction within that cone and this leads to uncertainty on the Δq_x, the x component of the pulse. If the photon impulse module is q and q_x its component along x, Δq_x is given by,

$$\delta q_x \approx q \sin \alpha. \tag{2.3}$$

Using the conservation of impulse in a collision between a photon and an electron to determine p at the time of observation, we see that it has to be,

$$\Delta p = \Delta q_x \tag{2.4}$$

Then using the de Broglie relation $q = 2\pi\hbar/\lambda$, the relation obtained is,

$$\Delta q \Delta p \approx 2\pi\hbar. \tag{2.5}$$

We observe that while Δx and Δp can be changed separately by playing on the parameters λ and α, their product is independent of these parameters. We must then consider additionally that in a real experiment there are other sources of error, so the uncertainty relationship takes the form of an inequality. If we analyze the physical origin of the Eq. (2.5), we realize that this is due to the dual nature of light, wave and corpuscular, on one hand lead to the indetermination of the position due to diffraction of the electron and the other hand an inaccurate localization of the trajectory of the photon. However, if we were to imagine a different experience that does not use the light as a means of observation, you would arrive at the same result, because the electron itself has a wave nature. The uncertainty relation is then intimate related to the corpuscular-wave particle duality.

According to quantum mechanics, a particle like an electron is described by a wave function, which is a complex function $\Psi(x, t)$ square integrable and normalized, such that $\int |\Psi(x,t)|^2 dx = 1$. The shape of the Ψ function, which is also called a wave packet, describes the wave aspect of the particle, while the corpuscular aspect is given by a statistical interpretation of the wave function. For example, if we wanted to measure the position of the particle with a detector we would do that at a very specific point x, while a priori we have a distribution of probability to find it anywhere, with a density of probability for unit volume given by $|\Psi(x,t)|^2$. In a simple case like the wave function we can derive the relation of indeterminacy. Let us take the following wave function, dependent on one space variable x:

$$f(x) = Ne^{-x^2/4a^2}, \tag{2.6}$$

where $N = (2\pi a^2)^{1/4}$ is a normalization constant. The probability density $|f(x)|^2$ is a Gaussian centered at the origin. Since the uncertainty Δx is the standard deviation of x, and square mean value is,

$$(\Delta x)^2 = \int_{-\infty}^{\infty} x^2 |f(x)|^2 dx = \frac{1}{\sqrt{2\pi}a} \int_{-\infty}^{\infty} x^2 e^{-x^2/2a^2} dx = a^2 \tag{2.7}$$

and then to $\Delta x = a$. The f(x) can be written in Fourier integral according to the formula

$$f(x) = \frac{1}{\sqrt{2\pi}} \int_{-\infty}^{\infty} F(k) e^{ikx} dk \tag{2.8}$$

This expresses the $f(x)$ as a linear combination of continuous waves with amplitude $F(k)$ and wave number $k = \lambda\pi/2$.

We can interpret $|F(k)|^2$ as the density of probability of finding the particle with wave number k, or, as regards the formula the de Broglie, with momentum $p = \hbar k$. The function $F(k)$ results in,

$$F(k) = \frac{1}{\sqrt{2\pi}} \int_{-\infty}^{\infty} f(x) e^{-ikx} dx = 2Nae^{-k^2 a^2}. \tag{2.9}$$

which is still a Gaussian centered around $k = 0$ and standard deviation,

$$\Delta k = \frac{1}{2a} \tag{2.10}$$

By (2.7) and (2.10) and the relation $\Delta p = \hbar \Delta k$ we finally obtain,

$$\delta x \Delta p = \frac{\hbar}{2} \tag{2.11}$$

So, it is proven that the Gaussian function minimizes the product of the indeterminacy, then for a generic wave function we will have to write

$$\delta x \Delta p \geq \frac{\hbar}{2}. \tag{2.12}$$

2.1.1.1 *Consequences of the Uncertainty Principle*

Given the smallness of the constant $\hbar$ ($\hbar = 1.05461 0^{27} ergs$), the effects of the uncertainty principle are negligible in the macroscopic world. It can be said that the uncertainty principle establishes an absolute distinction between systems "large" and "small" or micro-systems.

This principle also introduces the concept of incompatible sizes, a completely new concept compared to classical physics. It is said that two quantities are compatible if you can measure simultaneously with an arbitrary precision. The Eq. (2.12) then establish that q and p are not compatible. In particular a precise measure of q, ideally with $\Delta q = 0$, would leave the corresponding Δp to an indeterminacy, a very large value. This means that a measure p would give any result with the same probability. If q and p are not compatible, in general nor are two independent dynamic quantities $A(q, p)$ and $B(q, p)$.

In classical mechanics a physical quantity is called a dynamical variable if it is a function of the canonical coordinates q and p, $A = A(q, p)$. It is understood that the value of the function must match the measured result. It is clear that in quantum mechanics the relation $A = A(q, p)$ may not have the same meaning, and we can infer the value of A from the values of the coordinates q and p, since they cannot be determined at the same instant. The value of A can only be achieved by a direct measurement. It is said that a physical quantity is defined in an operating mode, meaning that it can be determined by using a specified measurement procedure.

Consider as an example the angular momentum of a point particle relative to the origin, defined by $L = x \times p$. The incompatibility between x and p produces the following consequences:

- The value of a component, for example, $L_z = x p_y y p_x$ may not be inferred from knowledge of the variables x, y, p_x, p_y, because they are not compatible.

- L_z itself is not compatible with any of the variables x, y, p_x, p_y.

- Two different components of L are not compatible with each other.

Another important consequence of the uncertainty principle relates to the dynamical equations. Consider for example the canonical equations of classical mechanics

$$\begin{cases} \dot{q}(t) = \frac{\partial H(q,p)}{\partial p_i} \\ \dot{p}(t) = -\frac{\partial H(q,p)}{\partial q_i} \end{cases} \tag{2.13}$$

These equations can be integrated, and we can then determine completely the motion of the system, given the initial conditions $q_i(0)$ and $p_i(0)$. But because of the uncertainty principle, it follows that the motion of the system is still not known, since these initial conditions are not completely determine.

It follows that the trajectory could not be exactly determined. The trajectory is defined as the curve traveled by the representative point of the space of configurations, with parametric equations $q_i = q_i(t)$. Classically the trajectory is uniquely determined by the initial conditions $q_i(0)$ and $\dot{q}_i(0)$, but again, being $\dot{q}_i(0)$ function of p_i, the initial conditions cannot be exactly determined and therefore the trajectory is indeterminate.

At the atomic level, this indeterminacy and the very concept of trajectory loses its meaning. It is well known that particles leave visible traces along their trajectory in some dedicated detectors such as photographic emulsions and bubble cloud chambers. We realize that the existence of these tracks is compatible with the uncertainty principle because of their finite width, ranging from about 1 m for photographic emulsions to approximately 1 mm for bubble chambers. If we call d the width of the track, the observation of the particle at a point on the trajectory results in an angular uncertainty equal to $\alpha] \approx \Delta p_\perp / p \approx h/pd = \lambda/d$ where λ is the de Broglie wavelength. If we consider a non-relativistic electron of 1KeV, $\lambda = 3.910^9$ cm and $d = 1$ m, then $\alpha = 3.910^5$, about 8 arc seconds, which is pretty much negligible. The value of α is even smaller for heavier particles or less energetic.

States, Sizes, and Measurements

Let us consider a mechanical system with N degrees of freedom. Classically a state of the system is completely determined if you know at any given moment the $2N$ canonical coordinates (q, p). As we said, a physical quantity is represented by a function of q and p: $A = A(q, p, t)$ and the value of the function at a given time is the result of a possible measure. Therefore, the measurement of a given size in a given system is always a unique result.

In quantum mechanics, the situation is fundamentally different, because the coordinates q and p are not compatible they are not enough to determine the state of the system. A state of the system, which we call a quantum state, will be determined from the knowledge of a set maximum of certain quantities that are independent and compatible with each other.

A measurable physical quantity is called an observable. If at a given moment a measurement is performed by one or more compatible observables it is said they are performing an observation of the system.

A briefly remark is an independent set of maximum and compatible observables. For example, the Lagrangian coordinates $q_1, ..., q_N$ and the conjugate momenta $p_1, ..., p_N$ constitute two different maximum observations of a system. This shows us that a generic maximum observation consists of General set of N variables, while in classical mechanics this number is $2N$.

Therefore, let it be $A_1, A_2, ..., A_N$ a "maximum set of observables" of a system and $a = a_1, a_2, ..., a_N$ the observed values of these. The corresponding quantum state will be labeled with the symbol $|a\rangle$, notation introduced by Dirac. If the system is in the State $|a\rangle$ and a measurement A_i is performed, one of a_i will result and is unique. In this case we will say that $|a\rangle$ is an eigenstate of A_i.

Every observable admits at least one eigenstate to every possible result of a measurement. Conversely, a given $|a\rangle$ of the system, if a measurement of an observable B incompatible with A will not give a unique result.

This means that if we repeated many times the measure of B, the system always in the $|a\rangle$ state, you would get different results each time $b_1, b_2, ...$ The result of a single measure is not predictable, but quantum mechanics allows us to calculate a priori the probability of getting different results.

In general the result of a measurement is notas in classical mechanicsa unique function of the State. An observation of the system affects it strongly, because it causes and abruptly change in the quantum state. In fact, if the system is initially in the $|a\rangle$ and a B measure is done, obtaining the result b_k, after the measurement the system will be in the eigenstate $|b_k\rangle$.

Superposition Principle

If a system is in the State $|a\rangle$ and B is measured, it has a certain probability is P_k to find the result b_k which is equivalent to finding the systemeigenstate $|b_k\rangle$, with $k = 1, 2, \dots$. We could express the State $|a\rangle$ as a statistical mixture of eigenstates $|b_k\rangle$ with weights P_k. This description is no necessary correct, because before making the measurement, the quantum state of the system is well defined, a pure state, and not a statistical mixture of states.

Quantum mechanics tell us that the State $|a\rangle$ is a superposition of eigenstates $|b_k\rangle$ base on the following principle:

A generic quantum state of a system can be think of as a superposition of eigenstates of an observable data.

The converse is also true, and can be stated in general form by saying that a superposition of pure States and still a pure State.

With the concept of "overlap" means that the System State is constituted at the same time by all its eigenstates components, each of which operates not only with the given weight, but consistently, and even with a data "phase."

Example: States of Polarization of a Photon

As an example of overlapping States, we discuss the case of the States of polarization of a photon. Consider a monochromatic linearly polarized beam of light, propagating along the z axis. $\hat{e}$ is the unit vector of polarization, lying in the xy plane, and we pass the beam through a polarizing filter, which in this case serves as a polarization analyzer. The filter is in the xy plane and can rotate about the axis z. It has a direction characteristic $\hat{e}_1$, orthogonal to the axis z, which is the direction of polarization of the light. The following phenomena are observed in classical optics:

- If $\hat{e}$ is parallel to $\hat{e}_1$ the beam passes unchanged through the filter, while if $\hat{e}$ and $\hat{e}_1$ are orthogonal, the beam is absorbed completely.

- If $\hat{e}$ forms an angle α with $\hat{e}_1$, the outgoing light comes polarized along $\hat{e}_1$ and has an reduced intensity compared to incident by a factor of $\cos \alpha^2$.

We observe that the first case correspond to a particular $\alpha = 0$ and $\alpha = \pi/2$ of the second case. According to quantum mechanics, the ray of light consists of a bundle of identical photons, each of which can be represented by a wave train with propagation vector $\vec{k}$ parallel to z, frequency $\omega = ck$ and polarization orthogonal to k. Because each photon acts as individually, we wonder what happens to a single photon when faces the polarizing filter. We take a system of orthogonal axes x and y, so that $\hat{e}_1$ is the unit vector along the x axis and $\hat{e}2$ the unit vector on the y axis. Our first case admit an obvious interpretation. If the incident photon polarization $\hat{e}$ is parallel to $\hat{e}_1$, then the photon passes unchanged, while if is parallel to $\hat{a}_2$ photon is absorbed.

In these cases the result of observation is unique, then $\hat{e}_1$ and $\hat{e}_2$ represent two eigenstates of the polarization of the photon.

The interpretation of the second case is not so evident. Because the incident photons are all the same, we could feel the naive impulse of thinking that everyone would face the same fate; but because a single photon cannot be divided into two, it must be admitted that some photons pass through the filter and others are absorbed. Thus, we cannot predict what will be in the fate of a single incident photon, but we can say that it has a "chance" of passing through equal to $cos\alpha^2$.

We also observe that the photons that pass through the filter change their polarization from $\hat{e}$ to $\hat{e}_1$. The superposition principle says in our case that the state of polarization $\hat{e}$ we can be thought in relation like an superposition of two polarization eigenstates with $\hat{e}_1$ and $\hat{e}_2$, with a probability $\cos \alpha^2$ and $\sin \alpha^2$, respectively. We can go one step further by saying that to the concept of "superposition" of states we can give the formal meaning of a linear combination, if we make the polarization states correspond to the corresponding unit direction. In fact if we break up the unit vector $\hat{e}$ along the x and y axes we have:

$$cos\alpha\hat{e}_1 + sin\alpha\hat{e}_2. \tag{2.14}$$

This result also shows that the probability of finding the polarization of the photon in one of the eigenstates $\hat{e}_1$ or $\hat{e}_2$ is given by the square of its coefficient.

Generalization of the Results

The polarization of light is particularly simple, since it has to do with two states only. However, the results have a validity to a general case, if it is properly formulated, and is part of the postulates of quantum mechanics. These results can be summarized as follows.

- We can associate to a generic quantum state an element $|\psi\rangle$ of a suitable vector space, which we call with Dirac a vectors or "ket." If this vector is normalize we call it a "vector state."

- Every observable magnitude admits a set of eigenstates $|a_k\rangle$ called eigenvectors, there are at least one for each possible result a_k of the measurement.

- The set of eigenvectors form an orthonormal basis, so any state vector can be expressed as a linear combination of eigenvectors:

$$|\psi\rangle = \sum_k c_k |a_k\rangle. \qquad (2.15)$$

- The probability P_k of finding the result a_k by measuring A is the modulus square of the coefficient c_k:

$$P_k = |c_k|^2. \qquad (2.16)$$

Let us consider the vector space of states. The size of the space and the number of independent states of the system. In the case of the polarization of a photon propagating in a fixed direction, there are only two independent states, allowing us to represent those states using unit vectors in the plane xy. But in general a system can have infinite states. Just think of the oscillation modes of the electromagnetic field in a cavity, which is a pulse of particle states confined in a slab-shaped region, or the energy levels of an electron in a hydrogen atom. In these cases the space of states have infinite dimension. Also the space must be endowed with a scalar product.

This allows us to obtain the coefficients c_k of (2.15) as the scalar product between the $|a_k\rangle$ and $|\psi\rangle$, that in the Dirac formalism is,

$$c_k = \langle a_k | \psi \rangle. \tag{2.17}$$

This allows us to calculate the probability P_k, which is solely dependent on the initial state and the values a_k. Finally, the superposition principle requires that any vector state $|\psi\rangle$ can be expanded according to (2.15). We ask that the converse is also true, and that any linear combination of convergent $|a_k\rangle$ corresponds to a vector state of the system. This property characterize a Hilbert space. We state as principle that the vectors states form a Hilbert space $\mathcal{H}$. From the previous discussions, it appears that in quantum mechanics, observables must have the following properties.

- To a given observable quantity A we can associate a set of values $a_1, a_2, ...$, that are the possible outcomes of measurements. A given result a_k is associated to an eigenvector $|a_k\rangle$.

- The measure on a generic state $|\psi\rangle$ do not provide an unique result, what we can obtain is an outcome with a given probability P_k.

- Two observables A and B are not in general compatible. In this case there can be no system states that are at the same time, eigenstates of A and B.

It is clear that the observable quantity A cannot be represented by a function of the characteristics variables of the state, since this would always give a unique result. To solve this, we can use the concept of self-adjoint linear operator in Hilbert space.

Denote $\hat{A}$ the operator representing the quantity A. We can write the eigenvalue equation,

$$\hat{A}|a_k\rangle = a_k|a_k\rangle, \tag{2.18}$$

There is a theorem which ensure us that the eigenvalues a_k are real and the eigenvectors $|a_k\rangle$ are orthogonal. This allows us to interpret the eigenvalues a_k as measurement results and the eigenvectors $|a_k\rangle$ as the corresponding eigenstates of the system. If we take all the $|a_k\rangle$, normalized, as an orthonormal basis of the space $\mathcal{H}$, a generic vector state $|\psi\rangle$ can be developed according to (2.15). If $|\psi\rangle$ is not one of the eigenvectors $|a_k\rangle$, any measure can give all results a_k in the expansion (2.15), with probability given by P_k. Between two operators $\hat{A}$ and $\hat{B}$ it is defined a product $\hat{A}\hat{B}$, which in general is not commutative: $\hat{A}\hat{B} \neq \hat{B}\hat{A}$. Two quantities are compatible if and only if the corresponding operators commute, that is $\hat{A}\hat{B} = \hat{B}\hat{A}$. It also shows that if and only if two operators commute, they admit a common set of eigenvectors.

Summary of the Mathematical Formalism

The physical system that we consider is associated with a Hilbert space $\mathcal{H}$. To indicate the vector state, scalar products and projectors we use Dirac vector notation. If $\hat{A} = \hat{A}\dagger$ is a self-adjoint linear operator that represents an observable of the system, and let be Σ and Λ the continuous and discrete spectrum respectively of $\hat{A}$. If $a_k \in \Lambda$ is an eigenvalue, we denote $|a_k, r\rangle$, an complete orthonormal set of eigenvectors, where $r = 1, ..., dk$ the number of independent degenerate eigenvectors. The space of dimension dk subtended by these eigenvectors is called eigenspace on a_k. Denoting with $\hat{P}_k$ the projector on this eigenspace, we have the following equations:

$$\hat{A}|a_k, r\rangle = a_k|a_k, r\rangle \tag{2.19}$$

$$\langle a_k, s|a_l, r\rangle = \delta_{kl}\delta sr \tag{2.20}$$

$$\sum_r |a_k, r\rangle\langle a_k, r| = \hat{P}_k \tag{2.21}$$

$$\hat{P}_k\hat{P}_l = \delta_{kl}\hat{P}_k \ ; \ Tr\hat{P}_k = dk. \tag{2.22}$$

If $a_k \in \Sigma$, there are no eigenvectors with finite norm. It is useful to introduce the generalized eigenvectors $|a, r\rangle$, such that:

$$\hat{A}|a, r\rangle = a|a, r\rangle \tag{2.23}$$

$$\langle a, r|a', s\rangle = \delta_{rs}\delta(aa') \tag{2.24}$$

$$\sum_r |a, r\rangle\langle a, r| = \frac{d}{da}\hat{P}(a_0, a), \quad a_0 \in \Sigma. \tag{2.25}$$

where $\delta(aa')$ is the Dirac delta distribution and $\hat{P}(a_0, a)$ the projector associated with the range $(a_0, a)\Sigma$, we can write it in the form,

$$\hat{P}(a_0, a) = \int_{a_0}^{a} \sum_r |a, r\rangle\langle a, r|da. \tag{2.26}$$

The Eq. (2.25) implies that the vectors $|a, r\rangle$ does not have finite norm, since $\langle a, r|a, r\rangle = \delta(0)$ which it is not finite. Therefore, it do not belong to $\mathcal{H}$, but to a larger space. For this reason the vectors $|a, r\rangle$ cannot represent physical states of the system, but can be a generalized basis in $\mathcal{H}$.

The completeness tell us that,

$$\sum_k \hat{P}_k + \int_\Sigma d\hat{P}(a_0, a) = \mathrm{I} \tag{2.27}$$

Obviously if the spectrum of $\hat{A}$ is only discreet, or continuous, only to the one of the terms in (2.27) appear. A generic vector on the Hilbert space can be written,

$$|\psi\rangle = \sum_{r,k}\langle a_k, r|\psi\rangle|a_k, r\rangle + \int_\Sigma \sum_r \langle a, r|\psi\rangle|a, r\rangle da \tag{2.28}$$

Applying the operator $\hat{A}$ to (2.25), performing the sum and the integral, we get:

$$\hat{A} = \sum_{r,k} a_k|a_k, r\rangle\langle a_k, r| + \int_\Sigma a \sum_r |a, r\rangle\langle a, r|da. \tag{2.29}$$

This expressions represents the spectral decomposition of the operator $\hat{A}$.

Quantum States and Their Representation

An important distinction that we have to do is between a pure state and statistical mixture of states.

A pure state is a well defined state from the quantum point of view, this is a State for which a given maximum observation $\hat{A} = \hat{A}(1), \hat{A}(2), ..., \hat{A}(n)$ gives a well-defined result consisting of a set of eigenvalues, all belonging to the discrete spectrum.

Examples of pure States are the eigenstates of a one-dimensional harmonic oscillator and angular momentum and energy eigenstates for bound states of the hydrogen atom. In a pure state you can associate a vector $|a\rangle$ in Hilbert space $\mathcal{H}$, which is a simultaneous eigenvector of $\hat{A}(i)$ with eigenvalues $a_{k_i}(i)$. However, this map may not be bijective, because the eigenvector $|a\rangle$ is determine modulo a multiplying factor. We can instead establish a correspondence between a pure state and the set of all vectors proportional to $|a\rangle$. This set is a one-dimensional subspace of $\mathcal{H}$, which belongs to the space of $\hat{A}(i)$ with eigenvalues $a_{k_i}(i)$. We can also restrict the set of normalized vectors $e^{i\alpha}|a\rangle$, with $\langle a|a \rangle = 1$ and $0 \leq \alpha <= 2$. In addition the pure states we may encounter the following cases.

- The system is not completely determined, and the observation made on the system not an maximal observation.

- Observable with continuous spectrum are measured, such as position or momentum of a particle. In this case, a measure of such observable $\hat{A}$ made with any real instrument, cannot give exactly a value $a \in \Sigma$, but will be affected by some experimental error.

- The system consists of a statistical ensemble of identical systems, for example, of atoms, which are in different quantum states. In this case the State of a generic system cannot be described by a pure state, but you are dealing with a statistical mixture of States.

The first case can be brought into the third by arbitrarily assigning the same "chance" to all values of the variables that are not measured, whereas case two can be traced back to a pure State in the limit where variations of the a are infinitesimal. Usually we speak of eigenstates of the energy even for the unbounded case, for which the values of these variables belong to the continuous. From a conceptual point of view, we have to remember that the ideal extrapolations of state cannot be physically realized. This extrapolation is reflected in the fact that the states have infinite norm.

Ultimately, the states of a system can be of two types: statistical mixture or pure. In fact, by the second part of the first postulate tells that any physical state can be represented by an operator $\hat{W}$, called a statistical operator or a density operator. The physical information needed to specify the state and therefore the operator $\hat{W}$, is given by the statistical distribution of the results of a given observation.

To define the statistical operator, we start from the pure state case, which you can think of as a special case of a mixture. We show that in a pure state has an associated projector $\hat{P}_\phi = |\phi\rangle\langle\phi|$, with $Tr\hat{P}_\phi = dim(h_\phi) = 1$, and such that $h_\phi = \hat{P}_\phi\mathcal{H}$. We can therefore a pure state can be associate with the projector $\hat{P}_\phi$ and we can identify $\hat{W}$ with $\hat{P}_\phi$:

$$\hat{W} = \hat{P}_\phi = |\phi\rangle\langle\phi|. \tag{2.30}$$

Consider now the case of a observable $\hat{A}$ with discrete spectrum. Suppose you have a collection of many identical systems, which can be found in different quantum states, and we are about to perform a measure of $\hat{A}$ on a generic system. If $a_k \in \Lambda$ the possible results of measurement are w_k or relative probability. Then $\hat{W}$ is given by,

$$\hat{W} = \sum_k w_k \hat{P}_k = \sum_k w_k |a_k\rangle\langle a_k|. \tag{2.31}$$

This expression extends (2.30) to the case of a statistical distribution of pure states. It has the same shape as the spectral representation for the discrete part of (2.29) and tells us that the eigenvalues of $\hat{W}$ are the probability w_k, such that,

$$0 \le w_k \le 1, \quad \sum_K w_k = 1. \tag{2.32}$$

From (2.31) and (2.32) follows that $\hat{W}$ is a bounded operator with norm $||\hat{W}|| \le 1$, self-adjoint, positive definite and trace equal to 1. Explicitly we have:

$$||\hat{W}|| \le 1; \quad \hat{W}^\dagger = \hat{W}; \quad \langle\hat{W}\rangle \ge 0; \quad Tr\hat{W} = 1. \tag{2.33}$$

We observe that the properties of the trace inn the case of a pure state corresponds, that the state vector is normalized. From (2.31), using the relation $Tr\hat{P}_k = dk$ with dk = 1, we get:

$$Tr\hat{W}^2 = \sum_k w_k^2 \le 1, \tag{2.34}$$

and it can be seen that only if $k = k_0$, $w_{k_0} = 1$, and $w_k = 0$ for any $k \ne k_0$, we are in the case of a pure State. Therefore, the condition of pure state is,

$$Tr\hat{W}^2 = Tr\hat{W} = 1. \tag{2.35}$$

Having finite trace, the operator $\hat{W}$ may not have a continuous spectrum. Therefore, if the observable $\hat{A}$ has a continuous spectrum with generalized eigenvectors $|a\rangle$, $\hat{W}$ may not be expressed in the form $\int_\Sigma g(a)|a\rangle\langle a|$ which would be the natural generalization of (2.31).

The matrix representing $\hat{W}$ in a particular basis is called a density matrix. Especially in the basis $|a_k\rangle$ from (2.31)

$$W_{kl} = \langle a|\hat{W}|a_k\rangle = w_k \delta_{kl}, \tag{2.36}$$

in this basis the matrix $\hat{W}$ is diagonal, with elements given by the probability w_k.

This is a special case of a general property that we can state as follow: $\hat{W}$ is the statistical operator of the system, $\hat{B}$ a given observable, b_k a generic eigenvalue of $|bk, r\rangle$ with $(r = 1, ..., d_k)$ the corresponding set of degenerate eigenvectors. Then, the probability that a measurement of $\hat{B}$ give the result b_k is given by the sum of the diagonal elements of the density matrix in the basis $|bk, r\rangle$:

$$w(b_k) = \sum_r \langle b_k, r|b_k, r\rangle, \tag{2.37}$$

where the sum over r is the sum over all the probabilities compatible with $\hat{B}$ which are not measured. If $\hat{B}$ has a continuous spectrum and $b \in \Sigma$, then:

$$w(b) = \sum_r \langle b, r|b, r\rangle, \tag{2.38}$$

where $w(b)$ represents the probability density for b.

The (2.37) and (2.38) extends immediately to the case that $\hat{B}$ is an observable compatible and $b_k(b)$ its corresponding set of eigenvalues.

From (2.37), by using the definition of the projector $\hat{P}(b_k)$ on the eigenspace of b_k, we can rewrite in the form

$$w(b_k) = Tr[\hat{P}(b_k)\hat{W}], \tag{2.39}$$

which is formally independent of a basis choice.

2.2 THE POSTULATES OF QUANTUM MECHANICS

The basic principles of quantum mechanics are expressed by a number of postulates. Their numbers and can vary from one text to another. We will formulate a list of seven distinct postulates, requiring only that they are independent from a conceptual point of view, although there may be enough natural relations among them, that is also suggested by the mathematical formalism.

2.2.1   Postulate 1: Quantum States

Given physical system S and associated Hilbert space $\mathcal{H}$. Each pure state of the system corresponds to vector of $\mathcal{H}$, which contains all the physical information about the system. We can also say, equivalently: a pure state of the system is represented by a vector $|\psi\rangle \in \mathcal{H}$, a vector state, normalized and determined modulo a phase factor. More generally: each quantum state of the system is a self adjoint operator $\hat{W}$, positive definite, bounded, linear and trace equal to 1, is said to be an statistical operator.

2.2.2   Postulate 2: Observables

Any observable on the system S is a self-adjoint linear operator $\hat{A}$ in the space $\mathcal{H}$.

2.2.3   Postulate 3: Measurement Values

The range of possible values for the measurements of an observable A and given by the spectrum of the operator of the corresponding $\hat{A}$.

2.2.4   Postulate 4: Probability of a Result

If the system is in a pure state represented by the normalized vector $|\psi\rangle$, the probability that a measurement of the observable $\hat{A}$ give the result $a_k \in \Lambda$ is given by,

$$w(a_k) = \langle\psi|\hat{P}(a_k)|\psi\rangle = \sum_r |\langle a_k, r|\psi\rangle|^2 \qquad (2.40)$$

where $\hat{P}(a_k)$ is the projector onto the eigenspace of $\hat{A}$ with eigenvalue a_k. The (2.40) is an extension of the basic equations (2.15) and (2.16) in the case of a degenerate space.

If the result of measurement of $\hat{A}$ results in a continuous spectrum, there is always some error with the form $\pm\delta$. In this case we have the probability that a measurement of $\hat{A}$ gives a result in the range $(a_1, a_2) \in \Sigma$ and given by,

$$w(a_1, a_2) = \langle\psi|\hat{P}(a_1, a_2)|\psi\rangle \int_{a_1}^{a_2} \sum_r |\langle a, r|\psi\rangle|^2 da, \tag{2.41}$$

where for the projector $\hat{P}(a_1, a_2)$ is used the expression (2.26). In the limit $a_2 a_1 = \delta a \to 0$, we can define the probability density for a unit range as follows:

$$w(a) = \lim_{\delta a \to 0} \left[\frac{1}{\delta a} w(a, a + \delta a) \right], \tag{2.42}$$

where we assume that $w(a)$ is a real non-negative continuous function of a. From (2.41) is clear that the probability density $w(a)$ is given by,

$$w(a) = \sum_r |\langle a, r|\psi\rangle|^2. \tag{2.43}$$

Is easy to verify that the previous expressions fulfill the requirements of a probability. Firstly, we immediately see that $w(a_k)$ and $w(a)$ are non-negative. In addition, the sum of all probability, including the integral on the continuous case gives:

$$\sum_k w(a_k) + \int_\Sigma w(a) da = \sum_{r,k} \langle\psi|a_k, r\rangle\angle a_k, r|\psi\rangle \tag{2.44}$$

$$+ \int_\Sigma \sum_r \langle\psi|a, r\rangle\langle a, r|\psi\rangle) da = \langle\psi|\psi\rangle = 1,$$

Now consider the more general case of a statistical mixture of states. If the system state is described by a statistical operator $\hat{W}$, the probability of measuring of one or more compatible observables $\hat{A}$ give a result $a_k \in \Lambda$ is given by,

$$w(a_k) = \sum_k \langle a_k, r|\hat{W}|a_k, r\rangle = Tr[\hat{P}(a_k)\hat{W}]. \tag{2.45}$$

In the continuous case, the probability of finding a value within a range over (a_1, a_2) it is obtained from (2.45) by replacing $\hat{P}(a_k)$ with $\hat{P}(a_1, a_2)$. The probability density in the interval of $a \in \Sigma$ is given by,

$$w(a) = \sum_r \angle a, r|\hat{W}|a, r\rangle. \tag{2.46}$$

2.2.5   Postulate 5: Reduction of the State

If the system is in a pure State represented by the normalized vector $|\psi_0\rangle$, and a measurement is performed of one or more compatible observables $\hat{A}$, getting the result $a_k \in \Lambda$, then, immediately after the measurement, the system will be in a pure State $|\psi_1\rangle$ given by,

$$|\psi_1\rangle = \frac{1}{\sqrt{N}}\hat{P}(a_k)|\psi_0\rangle = \frac{1}{\sqrt{N}}\sum_r \langle a_k, r|\psi_0\rangle|a_k, r\rangle, \qquad (2.47)$$

where $\hat{P}(a_k)$ is the projector onto the eigenspace of a_k and N is a normalization factor, whose form is $|N| = \langle\psi_0|\hat{P}(a_k)|\psi_0\rangle = w(a_k)$, while the phase remains arbitrary.

We note that the final vector state $|\psi_1\rangle$ is an eigenvector of $\hat{A}$ with eigenvalue a_k. In particular, if $\hat{A}$ is a maximal observable $|\psi_1\rangle$ coincides with $|a_k\rangle$, modulo a phase factor.

If the result of measurement in the continuous spectrum is in the range (a_1, a_2), the vector state after measurement is given by,

$$|\psi_1\rangle = \frac{1}{\sqrt{N}}\hat{P}(a_1, a_2)|\psi_0\rangle. \qquad (2.48)$$

We can see in this case that $|\psi_1\rangle$ still represents a pure state, but it is not an eigenvector of $\hat{A}$, but only an approximate eigenvector, the smaller $\delta a = a_2 a_1$, more approximate is the eigenvector.

More generally, suppose the system before observation is described by statistical operator $\hat{W}_0$. Then, if the measure gave the result $a_k \in \Lambda$, after measuring the statistical system and operator is given by,

$$\hat{W}_1 = \frac{1}{N}\hat{P}(a_k)\hat{W}_0\hat{P}(a_k), \qquad (2.49)$$

where $N = Tr[\hat{P}(a_k)\hat{W}] = w(a_k)$. If the result of measurement is in the range $(a_1, a_2) \in \Sigma$, then $\hat{W}_1$ is given by the Eq. (2.49) replacing the argument in $\hat{P}$ and w with (a_1, a_2).

2.2.6   Postulate 6: Equation of Motion

Suppose the system is in a pure state represented by the State vector $|\psi(t)\rangle$. In non-relativistic quantum mechanics one postulates that $|\psi(t)\rangle$ the obey the following equation of motion:

$$ih\frac{d}{dt}|\psi(t)\rangle = \hat{H}|\psi(t)\rangle \tag{2.50}$$

where $\hat{H}$ is a self-adjoint linear operator that corresponds, in classical mechanics, to the Hamiltonian of the system and therefore is called Hamiltonian operator, or simply the Hamiltonian. If the state of the system is described by a statistical operator $\hat{W}(t)$, that obeys the following equation:

$$ih\frac{d}{dt}\hat{W} = \left[\hat{H}, \hat{W}(t)\right] \tag{2.51}$$

Let's do some observations. First of all, the Eq. (2.50) can be justified by the following reasoning. Suppose that the system is isolated and consider the vector $|\psi(t)_E\rangle$ that represents a energy state, with a well defined energy E and time-independent. By virtue of the Planck relation $E = \hbar\omega$, if E is well-defined so it will be ω, and hence the state must be a periodic phenomenon (a wave or more generally a state vector) with frequency ω. You can therefore assume that $|\psi(t)_E\rangle$ depends on time according to,

$$|\psi(t)_E\rangle = e^{-i\omega t}|\psi_E(0)\rangle \tag{2.52}$$

Substituting this expression in (2.50) we obtain,

$$\hat{H}|\psi_E(0)\rangle = \hbar\omega|\psi_E(0)\rangle = E|\psi_E(0)\rangle, \tag{2.53}$$

which is the eigenvalue equation of $\hat{H}$. Therefore, the operator $\hat{H}$, whose spectrum energy values, represents the observable energies, which classically corresponds to the Hamiltonian. Conversely, if $\hat{H}$ represents the observable energy, the equations (2.53) ar valid because of Planck relation.

On the other hand, following the superposition principle, any State vector $|\psi(t)\rangle$ can be expanded in series of eigenvectors $|\psi_E(t)\rangle$, and then, from the linearity of $\hat{H}$, it follows that $|\psi(t)\rangle$ obeys the equation of motion (2.50).

Is natural to consider $\hat{H}$ as a self-adjoint operator in an isolated system, because $\hat{H}$ represents the observable energy. However, the (2.50) must also apply to non-isolated systems, where energy is not preserved. In General, the condition that $\hat{H}$ self-adjoint can be deduced from the vector state, if its normalized for any time, condition required by the first postulate. In fact, from the relation $\langle \psi(t)|\psi(t) \rangle = 1$, and using (2.50) we get,

$$\frac{d}{dt}\langle \psi(t)|\psi(t) \rangle = \frac{1}{i\hbar}\langle \psi(t)|\hat{H} - \hat{H}^\dagger|\psi(t) \rangle = 0 \qquad (2.54)$$

and because $|\psi(t) \rangle$ and arbitrary state, it follows $\hat{H} = \hat{H}^\dagger$.

2.2.7 Postulate 7: Correspondence Between Observable and Operators

Given a system with n degrees of freedom, the choice of a Lagrangian coordinate system $q_1, q_2, ..., q_n$ determines the quantization scheme. With a scheme fixed, all the canonical variables $q_i, p_i (i = 1, ..., n)$ are associated with self-adjoint operators $\hat{q}_i, \hat{p}_i$, which obey the following relations, called canonical commutators:

$$[\hat{q}_i, \hat{q}_j] = 0; \quad [\hat{p}_i, \hat{p}_j] = 0; \quad [\hat{q}_i \hat{p}_j] = i\hbar\delta_{ij}I. \qquad (2.55)$$

In addition to each variable in classical dynamics $A_{cl}(q_i, p_i)$ there is a self-adjoint operator associated $\hat{A} = A(\hat{q}_i, \hat{p}_i)$, which is a function of the operators $\hat{q}_i$ and $\hat{p}_i$ and depends explicitly on Planck's constant $\hbar$. Its dependence on $\hat{q}_i$ and $\hat{p}_i$ secured a suitable "quantization rule and must be sufficient to satisfy the correspondence principle:

$$A(\hat{q}_i, \hat{p}_i) \underset{\hbar \to 0}{\to} A_{cl}(q_i, p_i) \qquad (2.56)$$

The limit $\hbar \to 0$ is a formal limit, where $\hbar$ is treated as a variable set systematically to zero in all equations, including the commutation relations. In this limit the variables $\hat{q}_i$ and $\hat{p}_i$ switch all together and can be identified with the classical variables q_i and $p_i t$. This limit is considered as the classical limit of the theory.

The correspondence principle then establishes a correspondence between classical quantities $A_{cl}(q_i, p_i)$ and quantum observables $A(\hat{q}_i, \hat{p}_i)$, requiring that these functions differ only in terms of order $\hbar$. If there are no ambiguities in the ordering of the operators in $A(\hat{q}_i, \hat{p}_i)$, the two functions are said analogous. This requirement is called the principle of analogy.

2.3 UNCERTAINTY AND EXPECTATION VALUES

The fourth postulate tells that the distribution of probability gives all possible values of an observable, when the system is in a given quantum state. We can specifically calculate the two main parameters of the distribution, which are the quadratic mean and standard deviation. The average value of an observable $\hat{A}$, given that is an average calculated a priori, is called expectation value, and is indicated with the symbol $\langle \hat{A} \rangle$. From the classic definition of mean we have,

$$\langle \hat{A} \rangle = \sum_k a_k w(a_k) + \int_\Sigma a w(a) da. \tag{2.57}$$

If the system is in a pure state $|\psi\rangle$, using the probability formulas and the spectral representation, we get

$$\langle \hat{A} \rangle = \sum_{k,r} a_k \langle \psi | a_k, r \rangle \langle a_k, r | \psi \rangle + \int_\Sigma a \sum_r \langle \psi | a, r \rangle \langle a, r | \psi \rangle da \tag{2.58}$$

$$= \langle \psi | \hat{A} | \psi \rangle. \tag{2.59}$$

Otherwise, if the state is described by statistical operator $\hat{W}$, we get

$$\langle \hat{A} \rangle = \sum_k a_k Tr \left[\hat{P}(a_k) \hat{W} \right] + \int_\Sigma a Tr \left[\frac{d}{da} \hat{P}(a_0, a) \hat{W} \right] da \tag{2.60}$$

$$= Tr \left[\hat{A} \hat{W} \right]. \tag{2.61}$$

In particular, it is noted that the expressions (2.43) and (2.45) are nothing but the expectation values of its projectors. The standard deviation is a measure of the dispersion of measurements around the average value. In quantum mechanics, it is rather the uncertainty of the observable $\hat{A}$, and denoted as ΔA. According to the classical definition of standard deviation, ΔA and given by

$$\Delta A = \sqrt{\langle (\hat{A} - \langle \hat{A} \rangle)^2 \rangle} = \sqrt{\left(\hat{A}^2 - \langle \hat{A}^2 \rangle \right)} \tag{2.62}$$

We can see that $\Delta A \leq 0$, and $\Delta A = 0$ only if the system is in a pure state.

2.4   COMPATIBILITY CONDITION BETWEEN OBSERVABLE

We want to find the condition of compatibility between two independent observable represented by self-adjoint operators $\hat{A}$ and $\hat{B}$. This requires that the two quantities can be measured so that the result of the first measure is not altered by the second. In this way the results of the two measures can be used to characterize the state of the system. In other words, the state of the system after the measurement is represented by a simultaneous eigenvector, of $\hat{A}$ and $\hat{B}$. Considering the general case requires as a condition of compatibility that, if the measures of the two observables are made in succession, the final state of the system depends not only on the results of the measures, but not of the order in which they were performed.

Suppose that the system is in an arbitrary pure State $|psi\rangle$ and the measurements of $\hat{A}$ and $\hat{B}$ are performed, with values a and b, we indicated briefly that measuring a means two possible cases: $a = a_k$ if $a \in \Lambda$ or $a = a + \delta a$ if $a \in \Sigma$, and similarly for b. If we measure $\hat{A}$ first and then $\hat{B}$, the final state is obtained by applying twice equations the 5th postulate, and find

$$|\psi'\rangle = N'\hat{P}(b)\hat{P}(a)\psi\rangle, \qquad (2.63)$$

where N' and a normalization constant, $\hat{P}(a)$ indicates the projector on eigenspace of a_k or the projector that is associated with the range $(a\delta a, a + \delta a)$, whichever is applicable, and similarly for $\hat{P}(p)$. However, if the measure $\hat{B}$ is performed before, the final state will be,

$$|\psi''\rangle = N''\hat{P}(b)\hat{P}(a)\psi\rangle, \qquad (2.64)$$

But if $\hat{A}$ and $\hat{B}$ are compatible, the final state must be the same in both cases, then $|psi'\rangle$ and $\psi''\rangle$ have at most differ by a phase factor. Because the initial state $|\psi\rangle$ is arbitrary, we get the condition,

$$\hat{P}(a)\hat{P}(b) = \lambda\hat{P}(a)\hat{P}(b), \qquad (2.65)$$

where λ is a constant. Multiplying both sides of the equation by $\hat{P}(a)$ and $\hat{P}(b)$ properly, and using the idempotence of projectors, it follows that $\lambda = 1$, so the (2.65) becomes,

$$\left[\hat{P}(a), \hat{P}(b)\right] = 0. \tag{2.66}$$

This must hold regardless the values of a and b. Given the spectral decomposition implies that the operators $\hat{A}$ and $\hat{B}$ follows,

$$\left[\hat{A}, \hat{B}\right] = 0. \tag{2.67}$$

It is easily seen that the converse is also true, and that if $\left[\hat{A}, \hat{B}\right] = 0$, then the observables $\hat{A}$ and $\hat{B}$ are compatible. We can state the following,

Theorem. *The necessary and sufficient condition for two observables to be compatible is that they commute.*

2.5 RELATION OF INDETERMINACY

The uncertainty principle is not explicitly included among the postulates of quantum mechanics, it can be deduced from the postulates themselves by means of the mathematical formalism. We aim to prove that given two incompatible observables $\hat{A}$ and $\hat{B}$ the relation $\Delta A \Delta B > 0$ hold. We consider in particular the case that our system is in a pure state. We introduce for convenience the average as,

$$\begin{cases} \hat{A}_0 = \hat{A} - \langle \hat{A} \rangle \\ \hat{B}_0 = \hat{B} - \langle \hat{B} \rangle \end{cases} \tag{2.68}$$

We have then for ΔA and ΔB,

$$\begin{cases} (\Delta A)^2 = \langle \psi | \hat{A}_0^2 | \psi \rangle \\ (\Delta B)^2 = \langle \psi | \hat{B}_0^2 | \psi \rangle \end{cases} \tag{2.69}$$

Multiplying term to term we get,

$$(\Delta A)^2 (\Delta B)^2 = ||\hat{A}_0 |\psi\rangle||^2 ||\hat{B}_0 |\psi\rangle||^2 \geq |\langle | \hat{A}_0 \hat{B}_0 \tag{2.70}$$

$$= |\langle \psi \frac{1}{2} [\hat{A}_0, \hat{B}_0] + \frac{1}{2} \hat{A}_0, \hat{B}_0 |\psi\rangle|^2$$

$$= |\frac{i}{2} \langle \psi | - i[\hat{A}, \hat{B}] |\psi\rangle + \frac{1}{2} \langle \psi | \hat{A}_0, \hat{B}_0 |\psi\rangle$$

$$= \frac{1}{4} \langle -i[\hat{A}, \hat{B}] \rangle^2 + \frac{1}{4} \langle \hat{A}_0, \hat{B}_0 \rangle^2$$

$$\geq \frac{1}{4}\langle -i[\hat{A}, \hat{B}]\rangle^2$$

where we used the fact that $\hat{A}_0$ and $\hat{B}_0$ are self-adjoint, the Schwarz inequality and we introduced the anticommutator $\hat{A}_0, \hat{B}_0 \equiv \hat{A}_0\hat{B}_0 + \hat{B}_0\hat{A}_0$.

Ultimately from we obtain the indeterminacy relation,

$$\Delta A \Delta B \geq \frac{1}{2}\langle -i[\hat{A}, \hat{B}]\rangle. \tag{2.71}$$

Especially if $\hat{A} = \hat{q}$ and $\hat{B} = \hat{p}$ are canonical conjugate variables, and we arrive to the Heisenberg uncertainty relation,

$$\Delta q \Delta p \geq \frac{1}{2}\hbar. \tag{2.72}$$

If $\hat{A}$ and $\hat{B}$ are compatible, $[\hat{A}, \hat{B}] = 0$, from which we get $\Delta A \Delta B \geq 0$, so there is no relation between ΔA and ΔB, as expected.

2.6 EVOLUTION OPERATOR

The equation of motion (2.50) determines the evolution of a vector state as a function of time. Because it is a first order differential equation, its solution $|\psi(t)\rangle$ at time t is determined by the initial condition $|\psi(t_0)\rangle$. Therefore, must exist an operator $\hat{U}(t, t_0)$, that we call the evolution operator, such that,

$$|\psi(t)\rangle| = \hat{U}(t, t_0)|\psi(t_0)\rangle. \tag{2.73}$$

The operator $\hat{U}(t, t_0)$ must be unitary in order to preserve in time the norm of $|\psi(t)\rangle$. We must have,

$$\langle \psi(t)|\psi(t)\rangle = \langle \psi(t_0)|\hat{U}^d agger(t, t_0)\hat{U}(t, t_0)|\psi(t_0)\rangle \tag{2.74}$$
$$= \langle \psi(t_0)|\psi(t_0)\rangle = 1, \quad \forall |\psi(t_0)\rangle \in \mathcal{H}_S$$

from which follows,

$$\hat{U}\dagger(t, t_0)\hat{U}(t, t_0) = I. \tag{2.75}$$

The operator $\hat{U}(t, t_0)$ has the following properties:

$$\hat{U}(t_0, t) = U^1(t, t_0) \tag{2.76}$$

$$\hat{U}(t_0, t_0) = I \tag{2.77}$$

$$\hat{U}(t, t0) = \hat{U}(t, t_1)\hat{U}(t_1, t_0) \tag{2.78}$$

$$i\hbar\frac{\partial}{\partial t}\hat{U}(t, t_0) = \hat{H}\hat{U}(t, t_0). \tag{2.79}$$

The evolution is uniquely determined by the differential equation of first order (2.79) and the initial conditions. If the Hamiltonian $\hat{H}$ does not depend on time, $\hat{U}$ is given formally by,

$$\hat{U}(t, t_0) = e^{-\frac{i}{\hbar}\hat{H}(t-t_0)}, \tag{2.80}$$

where the exponential, as a function of $\hat{H}$, is defined using a power series expansion. It can be easily verify that the expression for $\hat{U}(t, t_0)$ satisfies all the properties.

If $\hat{H}$ depends on time, the expression for $\hat{U}(t, t_0)$ is more complex than a simple exponential function, because in general $\hat{H}(t)$ and $\hat{H}(t')$ do not commute for $t \neq t'$. The differential equation and the initial condition can be summarized in the integral equation,

$$\hat{U}(t, t_0) = I + \frac{i}{\hbar}\int_{t_0}^{t}\hat{H}(t')\hat{U}(t', t_0)dt', \tag{2.81}$$

which verifies directly. This can be formally resolved by the following an iterative procedure. As a first step we take right side of (2.81), and substitute in itself, integrating in a different time t'',

$$\hat{U}(t, t_0) = I + \frac{i}{\hbar}\int_{t_0}^{t}\hat{H}(t')dt' + \left(-\frac{i}{\hbar}\right)\int_{t_0}^{t}dt'\int_{t_0}^{t'}\hat{H}(t')\hat{H}(t'')\hat{U}(t, "t_0). \tag{2.82}$$

By iterating the process finally gets to the series,

$$\hat{U}(t, t_0) = I + \sum_{n=1}^{\infty}\left(\frac{i}{\hbar}\right)^n\int_{t_0}^{t}dt_1\int_{t_0}^{t_1}dt_2...\int_{t_0}^{t_{n-1}}dt_n\hat{H}(t_1)\hat{H}(t_2)...\hat{H}(t_n). \tag{2.83}$$

The nth term of the series can be rewritten in the form,

$$\left(-\frac{i}{\hbar}\right)^n\int_{t_0}^{t}dt_1\int_{t_0}^{t_1}dt_2...\int_{t_0}^{t_n}dt_n T\left[\hat{H}(t_1)\hat{H}(t_2)...\hat{H}(t_n)\right], \tag{2.84}$$

Here, we introduced the so-called product T-ordered of the operators $\hat{H}(t_i)$, where these are ordered in increasing time growing from left to right. In this way we can rewrite the series as follows,

$$\hat{U}(t, t_0) = T \exp\left[-\frac{i}{\hbar} \int_{t_0}^{t} \hat{H}(t')dt'\right]. \tag{2.85}$$

This is a formal expression, which is defined using the power series expansion of the exponential, where the term n-th is given by (2.84). We conclude by observing that, in the event that the operator a different times commute, the temporal ordering is superfluous and (2.85) is reduced to an ordinary exponential.

CHAPTER 3

THE SCHRÖDINGER EQUATION

3.1 INTRODUCTION

In 1926, the Austrian physicist Erwin Schrödinger derived an equation of waves from the Hamilton variational principle inspired by the analogy between Mechanics and Optics. This equation explained much of quantum phenomenology was known at that time. Although it was clear that this described the temporal evolution of the quantum state of a non-relativistic physical system, it was only a few days after its publication when the German physicist Max Born developed the probabilistic interpretation of the main object of the equation, the wave function, which is still in valid today, as it became part of the so-called Copenhagen interpretation of quantum mechanics. His ambitions in approaching the task were to find, in analogy with optics, the limit at which the trajectories of the particles could be described in a deterministic way. It came to propose an interpretation of the wave function as density of charge that did not fructify.

3.2 SYMMETRY TRANSFORMATIONS

Physical laws must be invariant under certain symmetries, represented by the transformations of the mathematical object that define these laws. In particular in quantum mechanics one can ask for the transformations of states and observables.

These transformations must fulfill what constitutes the essence of the so-called Wigner's theorem, that is, that the transformed observables must possess the same possible sets of eigenvalues as the old ones and that the transformations of the states must give the same probabilities. The temporal evolution of the states must always be given by unit operators that conserve scalar products.

Thus the expression of the transformed observables is given by,

$$\hat{A}' = \hat{U}\hat{A}\hat{U}^\dagger \tag{3.1}$$

If we focus our attention on the infinitesimal transformations:

$$\hat{U}(s) \approx I + s\left(\frac{dU}{ds}\right)\bigg|_{s=0} \tag{3.2}$$

Unitarity implies that the derivative will be pure imaginary,

$$\hat{U}\hat{U}^\dagger \approx I + s\left(\frac{dU}{ds} + \frac{dU^\dagger}{ds}\right)\bigg|_{s=0} = I \Rightarrow \frac{dU}{ds} = \frac{dU^\dagger}{ds}. \tag{3.3}$$

Then assuming $\hat{U}(0) = I$, these unit operators must have the form,

$$\hat{U}(s) = e^{iGs}, \qquad G = G^\dagger. \tag{3.4}$$

where G are called the generator of the transformation associated to the parameters s. In a multi-parametric case, we would have relations,

$$\hat{U}(s) = \prod e^{is_\alpha G_\alpha} \simeq I + is_\alpha G_\alpha. \tag{3.5}$$

These generators will obey a series of commutation relations. If we consider two transformations and their inverse,

$$e^{isG_\mu} e^{isG_\nu} e^{-isG_\mu} e^{isG_\nu} \simeq I + s^2 [G_\mu, G_\nu]. \tag{3.6}$$

And taking into account that any composition of transformations must equal, except for phase, another transformation of the group, we have, observing that:

$$e^{i\omega}\hat{U}(s) = e^{i\omega} \prod_\alpha e^{is_\alpha G_\alpha} \simeq I + is_\alpha G_\alpha + i\omega I. \tag{3.7}$$

and the operator identity,

$$[G_\mu, G_\nu] = ic^\lambda_{\mu\nu} G_\lambda + ib_{\mu\nu} I. \tag{3.8}$$

Where the last term can be made zero by doing = 0.

3.3 EQUATION OF MOTION OF A FREE PARTICLE

In the case of the dynamics of a free particle, assuming non-relativistic velocities, the following unit operators are associated with each symmetry,

$$\hat{R}_\alpha(\theta_\alpha) \to e^{-i\theta_\alpha I_\alpha} \tag{3.9}$$

$$x_\alpha + a_\alpha \to e^{-ia_\alpha P_\alpha} \tag{3.10}$$

$$x_\alpha + v_\alpha t \to e^{iv_\alpha G_\alpha} \tag{3.11}$$

$$t + s \to e^{is\hat{H}} \tag{3.12}$$

To find the equations of motion we will ask for the temporal evolution,

$$t \to t' = t + s$$

$$|\psi(t)\rangle \to e^{is\hat{H}}|\psi(t)\rangle = |\psi(t - s)\rangle \tag{3.13}$$

That is, by doing $s = t$,

$$|\psi(t)\rangle = e^{-it\hat{H}}|\psi(0)\rangle = |\psi(t - s)\rangle \Rightarrow \frac{d}{dt}|\psi(t)\rangle = -i|\psi(t)\rangle. \tag{3.14}$$

which will constitute the equation of motion of the wave function. We used $\hbar = 1$. where the generator $\hat{H}$ is the Hamiltonian of the system.

3.4 TEMPORAL EVOLUTION PICTURES

Let us picture the evolution equation for the state operator. First, observe that the evolution of the wave function from an initial instant t_0 is given by,

$$|\psi(t)\rangle = \hat{U}(t, t_0)|\psi(0)\rangle \tag{3.15}$$

This evolution operator must satisfy the equation of motion,

$$i\frac{\partial \hat{U}(t,t_0)}{\partial t} = \hat{H}(t)\hat{U}(t,t_0) \Rightarrow \hat{U}(t,t_0) = I - \frac{i}{\hbar}\int_{t_0}^{t}\hat{H}(t')\hat{U}(t',t_0)dt'$$

$$(3.16)$$

where we use that $\hat{U}(t_0) = I$.

If $\hat{H}$ is the system is conservative the evolution operator will have the form,

$$\hat{U} = e^{-i\frac{\hat{H}(t-t_0)}{\hbar}}.$$

$$(3.17)$$

Otherwise it will not have a specific form and will depend on the case to be studied. Let's see what happens with state operators. Suppose we do not lose generality when we consider an operator representing a pure state:

$$\begin{aligned}
\rho(t) &= |\psi(t)\rangle\langle\psi(t)| \\
&= U(t,t_0)|\psi(t_0)\rangle\langle\psi(t_0)|U^{\dagger}(t,t_0) \\
&= U(t,t_0)\rho(t_0)U^{\dagger}(t,t_0)
\end{aligned}$$

$$(3.18)$$

Deriving with respect to time, and using the equation of motion for U we have:

$$i\hbar\frac{d\rho}{dt} = [H(t),\rho(t)].$$

$$(3.19)$$

If we center our attention to the temporal evolution of the mean values of the observables,

$$\langle A\rangle(t) = \text{Tr}\,[\rho(t)A]$$

$$(3.20)$$

we see that until now, we have assumed that the explicit temporal dependence is carried by the state, this is the so-called Schrödinger picture, noting that,

$$\langle A\rangle(t) = \text{Tr}\left[U(t,t_0)\rho(t_0)U^{\dagger}(t,t_0)A\right] = \text{Tr}\left[\rho(t_0)U^{\dagger}(t,t_0)AU(t,t_0)\right]$$

$$(3.21)$$

We can suppose that the temporary dependence is carried by observable operators, defining:

$$A_H(t) = U^{\dagger}(t,t_0)AU(t,t_0) \quad \rightarrow \quad \langle A\rangle(t) = \text{Tr}\,[\rho(t_0)A_H(t)] \quad (3.22)$$

This constitutes the so-called Heisenberg picture, where the dynamic variables are those that evolve.

In the case of a pure state, the mean values have the form, in the image of Schrödinger and of Heisenberg respectively,

$$\langle A \rangle(t) = \langle \psi(t)|A|\psi(t) \rangle \tag{3.23}$$

$$\langle A \rangle(t) = \langle \psi(t_0)|A_H(t)|\psi(t_0) \rangle \tag{3.24}$$

We can deduce the evolution equations of the observable in the Heisenberg picture, assuming that the Hamiltonian also evolves in that picture,

$$\begin{aligned}
\frac{dA_H(t)}{dt} &= \frac{\partial U^\dagger}{\partial t}AU + U^\dagger A\frac{\partial U}{\partial t} + U^\dagger\frac{\partial A}{\partial t}U \\
&= \frac{i}{\hbar}U^\dagger HAU - \frac{i}{\hbar}U^\dagger AHU + U^\dagger\frac{\partial A}{\partial t}U \\
&= \frac{i}{\hbar}\left[H_H(t), A_H(t)\right] + \left(\frac{\partial A}{\partial t}\right)_H \tag{3.25}
\end{aligned}$$

In this picture the density state operator will only account for the initial data of the preparation of the experiment. The evolution of the average will therefore be,

$$\frac{d}{dt}\langle A \rangle(t) = \text{Tr}\left[\frac{i}{\hbar}\rho(t_0)\left[H_H(t), A_H(t)\right] + \rho(t_0)\left(\frac{\partial A}{\partial t}\right)_H\right] \tag{3.26}$$

in the Schrödinger picture, we will have the similar formula,

$$\frac{d}{dt}\langle A \rangle(t) = \text{Tr}\left[\frac{i}{\hbar}\rho(t)\left[H, A\right] + \rho(t)\frac{\partial A}{\partial t}\right] \tag{3.27}$$

In the case of a pure state the general formula for the evolution of an expectation value is given by,

$$i\hbar\frac{d}{dt}\langle \psi(t)|A(t)|\psi(t) \rangle = \langle \psi(t)|\left[A(t), H(t)\right]|\psi(t) \rangle + i\hbar\langle \psi(t)|\frac{dA(t)}{dt}|\psi(t) \rangle \tag{3.28}$$

which indicates that for an arbitrary system the evolution of the expectation values of an observable is due in part to the evolution of the state vector and partly to the explicit dependence in time of the observable.

From here it can also be seen that a constant of motion is an observable that does not explicitly depend on time and that commutes with H,

$$A \text{ is conserved} \quad \Leftrightarrow \quad \begin{cases} \frac{\partial A}{\partial A} = 0 \\ [A, H] = 0 \end{cases} \tag{3.29}$$

from which it is deduced that for conservative systems the Hamiltonian will be a constant of the motion too.

3.4.1 Position Representation

Assuming that the space is continuous and that the three components of the position vector $x = (x_1, x_2, x_3)$ of the particle are kinematically indepen-dent, we can assume that the operators representing them commute, which is equivalent to saying they have a common set of eigenvectors, so that,

$$X_i |x\rangle = x_i |x\rangle \tag{3.30}$$

The representation of position or configuration of Hilbert space that we study, will be where the vectors are expanded in the orthonormal base of the position, i.e.,

$$|\psi\rangle = \int_V dx \, \langle x|\psi\rangle |x\rangle \tag{3.31}$$

With the usual orthonormality and closure relations for continuous spaces,

$$\langle x|x'\rangle = \delta(x - x') \, , \quad \int_V dx \, |x\rangle\langle x| = I \tag{3.32}$$

The coefficients in the expansion (3.31) define functions of continuous variable are identified with the wave function,

$$\psi(x) = \langle x|\psi\rangle \tag{3.33}$$

Operators act simply in Hilbert space or the functional space indifferently. To see the form that has the moment operator in this representation, we must remember its role as generator of translations,

$$e^{-\frac{i}{\hbar}aP}|x\rangle = |x + a\rangle \tag{3.34}$$

So that, for infinitesimal translations

$$\langle x + a|\psi\rangle = \langle x|e^{\frac{i}{\hbar}aP}|\psi\rangle \simeq \langle x|I + \frac{iaP}{\hbar}|\psi\rangle \tag{3.35}$$

in which the functional space takes the form

$$\psi(x + a) \simeq \psi(x) + \frac{iaP}{\hbar}\psi(x) \tag{3.36}$$

Comparing it with the Taylor series we have,

$$P = -i\hbar\nabla \tag{3.37}$$

Whose components in Cartesian coordinates have the simple form

$$P_x = -i\hbar\frac{\partial}{\partial x} \ , \quad P_y = -i\hbar\frac{\partial}{\partial y} \ , \quad P_z = -i\hbar\frac{\partial}{\partial z} \tag{3.38}$$

The Schrödinger wave equation for a particle subjected to a scalar potential $V(x)$ will be given by the expression of the non-relativistic Hamiltonian, which we assume to be, in the position representation,

$$H = \frac{1}{2m}P^2 + V(x) = -\frac{\hbar^2}{2m}\nabla^2 + V(x) \tag{3.39}$$

And so we have

$$\left(-\frac{\hbar^2}{2m}\nabla^2 + V(x)\right)\psi(x,t) = i\hbar\frac{\partial}{\partial t}\psi(x,t) \tag{3.40}$$

3.4.2 Continuity Equation

One of the observations to be made about the Schrödinger equation is that it preserves the norm of the state vector,

$$\frac{d}{dt}\langle\psi|\psi\rangle = \left(\frac{d}{dt}\langle\psi|\right)|\psi\rangle + \langle\psi|\frac{d}{dt}|\psi\rangle = -\frac{1}{i\hbar}\langle\psi|H^\dagger|\psi\rangle + \frac{1}{i\hbar}\langle\psi|H|\psi\rangle = 0 \tag{3.41}$$

which is an equivalent way of expressing the unitarity of the evolution or the Hamiltonian Hermiticity.

Now we are going to go a bit further, and we will see how the probability evolves, and to make sure that we do not find sources or sinks of probability in our domain of definition. For this we define the so-called probability density as,

$$|\langle x|\psi\rangle|^2 = \langle\psi|x\rangle\langle x|\psi\rangle = \psi^*(x,t)\,\psi(x,t) = |\psi(x,t)|^2 \qquad (3.42)$$

if we integrate it on a given volume, would give us the probability that the particle will be in that volume at time t. Let us see how it evolves in two instants of time,

$$
\begin{aligned}
\int_V dx\,|\psi(x,t_0)|^2 &= \int_V dx\,\psi^*(x,t)\,\psi(x,t) \\
&= \int_V dx\,\langle\psi(t_0)|x\rangle\langle x|\psi(t_0)\rangle \\
&= \langle\psi(t_0)|\psi(t_0)\rangle \\
&= \langle\psi(t)|\psi(t)\rangle \\
&= \int_V dx\,|\psi(x,t)|^2 \qquad (3.43)
\end{aligned}
$$

where we use conservation of the norm. On the other hand we can write, for an arbitrary volume Ω,

$$
\begin{aligned}
\frac{\partial}{\partial t}\int_V dx\,\psi^*\psi &= \int_V dx\left(\psi^*\frac{\partial\psi}{\partial t} + \psi\frac{\partial\psi^*}{\partial t}\right) \\
&= \frac{i\hbar}{2m}\int_V dx\left(\psi^*\nabla^2\psi - \psi\nabla^2\psi^*\right) \\
&= \frac{i\hbar}{2m}\int_V dx\,\nabla\cdot\left(\psi^*\nabla\psi - \psi\nabla\psi^*\right) \\
&= -\int_V \nabla\cdot\vec{J}\,dx = -\oint_S \vec{n}\cdot\vec{J}\,ds \qquad (3.44)
\end{aligned}
$$

Where we used the Gauss divergence theorem, n is the normal unit vector pointing outward Ω, and the so-called probability current density is,

$$\vec{J}(x,t) \equiv -\frac{i\hbar}{2m}\left(\psi^*\nabla\psi - \psi\nabla\psi^*\right) = \frac{\hbar}{m}\mathrm{Im}\left(\psi^*\nabla\psi\right) \qquad (3.45)$$

Since the region Ω is arbitrary, we arrive at the continuity equation,

$$\frac{\partial}{\partial t}|\psi(x,t)|^2 + \nabla\cdot\vec{J}(x,t) = 0 \qquad (3.46)$$

Formally identical to the equation of a fluid without sources or sinks, that is, probability propagates as a fluid.

Logically, if we choose the integration volume so that it coincides with the entire three-dimensional space, following the conditions norm to be finite, the wave function will be zero on the surface at infinity, the second term of the continuity equation will be annulled and recover the conservation of the norm itself. In most cases this in fact happens even if we restrict our volume to the domain problem.

3.5 SOLUTIONS OF THE SCHRÖDINGER EQUATION

To find solutions of the equation, the method of separating variables is usually tested, which works whenever the potential energy does not depend on time, and consists in assuming the wave function to be factorized in the form,

$$\psi(x, t) = \psi(x)\phi(t) \qquad (3.47)$$

So that this function must be of the form,

$$\psi(x, t) = \psi(x)e^{-\frac{i}{\hbar}Et} = \psi(x)e^{-i\omega t} \qquad (3.48)$$

where E is the total energy of the particle, ω is the angular frequency of the associated wave and the spatial part must satisfy the Schrödinger equation independent of time,

$$-\frac{\hbar^2}{2m}\frac{d^2\psi(x)}{dx^2} + V(x)\psi(x) = E\psi(x) \qquad (3.49)$$

3.5.1 Zero Potential: Free Particle

In this case the equation is reduced to the equation,

$$\frac{d^2\psi(x)}{dx^2} = -\left(\frac{p}{\hbar}\right)^2 \psi(x) = -k^2\psi(x) \qquad (3.50)$$

where k is the wave number, inverse of the wavelength, which is deduced from the de Broglie relation $p = \hbar k$, and the solutions will be of the form,

$$\psi(x) = e^{ikx} \qquad (3.51)$$

So that a particular solution of the Schrödinger equation will be the monochromatic plane wave,

$$\psi(x, t) = Ae^{i(kx - \omega t)} \tag{3.52}$$

without forgetting that the frequency depends on the wavelength by the dispersion ratio of the free particle,

$$\omega(k) = \frac{\hbar k^2}{2m} \tag{3.53}$$

This solution describes a traveling wave since, for example, the nodes of its real part lie at the points $x = (n + 1/2)\pi/k + \omega t/k$, which are displaced over time.

The probability density in this case does not give us information since the amplitudes of the waves are the same in all the spatial regions:

$$|\psi^*\psi| = |A|^2, \quad \forall x \tag{3.54}$$

indicating that the particle will have the same probability at any point in space. We can calculate the mean value of the moment of the particle, which will be given by,

$$\begin{aligned}
\langle P \rangle_{|\psi\rangle} &= \langle \psi | P | \psi \rangle \\
&= \iint dx' dx \, \psi^*(x', t) \langle x' | \left(-i\hbar \frac{\partial}{\partial x} \psi(x, t) \right) | x \rangle \\
&= \hbar k \iint dx' dx \, \psi^*(x', t) \psi(x, t) \langle x' | x \rangle \\
&= \hbar k \int dx \, |\psi(x, t)|^2 \ = \ p = \ \sqrt{2mE}
\end{aligned} \tag{3.55}$$

where we have used Dirac delta properties of integration and the normalization condition.

However, the free particle solution of wave equation is not normalizable throughout the space, it does not belong to the Hilbert space of square integrable functions and would not hold the postulates of quantum mechanics. This is an old issue in the Physics of ideal systems, not just quantum systems, which sometimes give rise to mathematical inconsistencies.

To begin with, it does not make much sense to consider all Euclidean space as the laboratory of our particle. In addition, a much more realistic sense of the problem is to consider it as a limiting process of a true physical phenomenon described by a wave packet, i.e., a group of traveling waves. In fact, it is very risky to ensure that the particle is in a state with a perfectly defined linear momentum, it is closer to reality assume a distribution of moments. In this way the corresponding probability density will be more close to the description of the movement of the free particle.

In any case, to get a better idea of what we are talking about, we can give an approximate description of the behavior of a wave packet by assuming a distribution in wave numbers due only to two wave components,

$$\psi(x,t) = \psi_1(x,t) + \psi_2(x,t)$$
$$\text{with} \quad \begin{cases} \psi_1(x,t) = \cos(kx - \omega t) \\ \psi_2(x,t) = \cos[(k + dk)x - (\omega + d\omega)t] \end{cases} \tag{3.56}$$

Using the trigonometric identity,

$$\cos\alpha + \cos\beta = 2\cos\left(\frac{\alpha + \beta}{2}\right)\cos\left(\frac{\alpha - \beta}{2}\right) \tag{3.57}$$

we find the expression, for $dk \ll 2k$ and $d\omega \ll 2k$:

$$\psi(x,t) \simeq 2\cos\left(\frac{dk}{2}x - \frac{d\omega}{2}t\right)\cos\left(kx - \omega t\right) \tag{3.58}$$

We see that the oscillations of ψ fall within a envelope whose amplitude varies periodically. The waves interfere and are reinforced alternately. The velocity of the individual waves can be evaluated considering the nodes of the second factor,

$$x_n = \frac{\pi}{k}\left(n + \frac{1}{2}\right) + \frac{\omega}{k}t \quad \rightarrow \quad v_1 = \frac{\omega}{k} = \frac{E/\hbar}{p/\hbar} = \frac{p}{2m} = \frac{1}{2}v \tag{3.59}$$

We remark that this wave velocity is not the particle velocity. The velocity that must have correspondence with the particle is the so-called group velocity, which is evaluated considering the nodes of the first factor of ψ,

$$x_n = \frac{\pi}{dk}(2n + 1) + \frac{d\omega}{dk}t \quad \rightarrow \quad v_2 = \frac{d\omega}{dk} = \frac{dE}{dp} = v \tag{3.60}$$

These considerations about the motion of the particle can be continued by calculating the average values of the observables and recovering the laws of Classical Mechanics, at least formally, which constitutes the content of the Ehrenfest theorem.

3.5.2   The One-Dimensional Harmonic Oscillator

The first realistic case of a potential is the harmonic oscillator. This potential is of enormous importance, since its solution can be obtained analytically and is a model for many physical systems, from the study of the vibrations of atoms in diatomic molecules to the vibrations of electromagnetic waves in quantum field theory.

For small vibrations choosing the axes appropriately, the potential to study has the parabolic form,

$$V(x) = \frac{1}{2}kx^2 \tag{3.61}$$

where k is the restitution constant, since a particle that moves on its influence experience a linear restoring force $F = -kx$.

3.5.2.1   Analytical Method

The time-independent Schrödinger equation have the form,

$$-\frac{\hbar^2}{2m}\frac{d^2\psi(x)}{dx^2} + \frac{1}{2}kx^2\psi(x) = E\psi(x) \tag{3.62}$$

Taking into account the expression of the classical frequency oscillation,

$$\omega = \sqrt{\frac{k}{m}} \tag{3.63}$$

The equation results,

$$\frac{d^2\psi}{dx^2} + \left[\frac{2mE}{\hbar^2} - \left(\frac{\omega m}{\hbar}\right)^2 x^2\right]\psi = 0 \tag{3.64}$$

For convenience we introduce the parameters,

$$\alpha \equiv \sqrt{\frac{\omega m}{\hbar}}\ , \quad \beta \equiv \frac{2mE}{\hbar^2} \tag{3.65}$$

then Eq. (3.64) is written,

$$\frac{d^2\psi}{dx^2} + \left[\beta - \alpha^4 x^2\right]\psi = 0 \tag{3.66}$$

Rescaling x properly, can be put,

$$u \equiv \alpha x \quad \rightarrow \quad \frac{d^2\psi}{du^2} + \left[\frac{\beta}{\alpha^2} - u^2\right]\psi = 0 \tag{3.67}$$

which for u very large takes the form,

$$\frac{d^2\psi}{du^2} = u^2\psi \quad \rightarrow \quad \psi = Ae^{-\frac{1}{2}u^2} + Be^{\frac{1}{2}u^2} \tag{3.68}$$

since the wave function must converge in infinity, we have,

$$\psi(u) = Ae^{-\frac{1}{2}u^2}, \quad |u| \rightarrow \infty \tag{3.69}$$

This suggests that general solutions sought are,

$$\psi(u) = AH(u)e^{-\frac{1}{2}u^2} \tag{3.70}$$

So that $H(u)$ does not explode at infinity. Substituting for the function $H(u)$,

$$\frac{d^2H}{du^2} - 2u\frac{dH}{du} + \left[\frac{\beta}{\alpha^2} - 1\right]H = 0 \tag{3.71}$$

This equation, unlike the original, can be solved by the power series technique. Identifying coefficients and obtaining recurrence relations between them, we obtain the solution in terms of two series,

$$H(u) = a_0 + a_1 u + \cdots \quad \rightarrow \quad a_{l+2} = -\frac{(\beta/\alpha^2 - 1 - 2l)}{(l+1)(l+2)}a_l \tag{3.72}$$

$$H(u) = a_0\left(1 + \frac{a_2}{a_0}u^2 + \frac{a_4}{a_2}\frac{a_2}{a_0}u^4 + \cdots\right) + a_1\left(u + \frac{a_3}{a_1}u^3 + \frac{a_5}{a_3}u^5 + \cdots\right) \tag{3.73}$$

It is clear that for u large the low powers terms in u contribute less, but the ratios of the high power terms of the series are analogous to those of the quadratic exponential,

$$\frac{a_{l+2}}{a_l} \simeq \frac{2}{l} \tag{3.74}$$

$$e^{u^2} = 1 + u^2 + \frac{u^4}{2!} + \cdots + \frac{u^l}{(l/2)!} + \frac{u^{l+1}}{(l/2+1)!} + \cdots \quad \rightarrow \quad \frac{\frac{1}{(l/2+1)!}}{\frac{1}{(l/2)!}} \simeq \frac{2}{l}$$

$$\text{(3.75)}$$

Since this function does not converge, we must cut the series at some value of $l = n$, defined by,

$$\frac{\beta}{\alpha^2} - 1 - 2n = 0 \quad \rightarrow \quad \frac{\beta}{\alpha^2} = 2n+1 \quad \rightarrow \quad \begin{cases} n = 1, 3, 5, \cdots & \text{if} \quad a_0 = 0 \\ n = 0, 2, 4, \cdots & \text{if} \quad a_1 = 0 \end{cases}$$

$$\text{(3.76)}$$

This solution defines polynomials called, Hermite polynomials,

$$\psi_n(u) = A_n e^{-\frac{1}{2}u^2} H_n(u) \tag{3.77}$$

$$\psi_n(x) = A_n e^{-\frac{1}{2}\alpha^2 x^2} H_n(\alpha x) \tag{3.78}$$

Where quantization rule can be set as,

$$\alpha = \sqrt{\frac{m\omega}{\hbar}}, \quad \beta = \frac{2mE}{\hbar^2} \quad \rightarrow \quad E = \left(n + \frac{1}{2}\right)\hbar\omega \tag{3.79}$$

where $n = 0, 1, 2, \cdots$.

The functions $\psi_{2n}(x)$ are even and the $\psi_{2n+1}(x)$ are odd. The normalization constant takes the form,

$$A_n = \sqrt{\frac{\alpha}{\sqrt{\pi}2^n n!}} \tag{3.80}$$

Taking into account that the first Hermite polynomials are,

$$H_0(x) = 1, \quad H_1(x) = 2x, \quad H_2(x) = 2 - 4x^2, \quad \cdots \text{ etc} \tag{3.81}$$

We can plot a graphic representation of the first eigenstates.

The energy levels are analogous to those deduced by Planck with the old quantum theory, except the zero point energy, which is $\hbar\omega/2$.

3.5.3 Algebraic Method

This method begins defining the operators,

$$a \equiv \frac{1}{\sqrt{2}}\left(u + \frac{d}{du}\right) \tag{3.82}$$

$$a^\dagger \equiv \frac{1}{\sqrt{2}} \left(u + \frac{d}{du} \right) \qquad (3.83)$$

$$N = a^\dagger a \qquad (3.84)$$

where the latter is formally Hermitian.

As can be seen the expression in terms of the dynamic variables and the Hamiltonian is,

$$H = \frac{P^2}{2m} + \frac{1}{2}kX^2 \quad \rightarrow \quad H = \frac{1}{2}\hbar\omega \left(-\frac{d^2}{du^2} + u^2 \right) \qquad (3.85)$$

and therefore,

$$H = \hbar\omega \left(a^\dagger a + \frac{1}{2} \right) = \hbar\omega \left(N + \frac{1}{2} \right) \qquad (3.86)$$

From which also the expressions are deduced,

$$a = \frac{1}{\sqrt{2}} \left(\alpha X + \frac{i}{\hbar\alpha}P \right) , \quad a^\dagger = \frac{1}{\sqrt{2}} \left(\alpha X - \frac{i}{\hbar\alpha}P \right) \qquad (3.87)$$

$$X = \frac{1}{\sqrt{2}\alpha} \left(a^\dagger + a \right) , \quad P = \frac{i\alpha\hbar}{\sqrt{2}} \left(a^\dagger - a \right) \qquad (3.88)$$

Let us see that in effect, the eigenvalues of the operator N are nonnegative integers, and this is why it is known as number operator. Let us consider the commutators,

$$\left[a, a^\dagger \right] = I , \quad [N, a] = -a , \quad \left[N, a^\dagger \right] = a^\dagger \qquad (3.89)$$

And suppose that $|n\rangle$ is the normalized eigenvector of the operator N with eigenvalue n, then,

$$Na|n\rangle = ([N, a] + aN) |n\rangle = (n-1)a|n\rangle \qquad (3.90)$$

$$Na^\dagger|n\rangle = \left(\left[N, a^\dagger \right] + a^\dagger N \right) |n\rangle = (n+1)a^\dagger|n\rangle \qquad (3.91)$$

That is, $a|n\rangle$ and $a^\dagger|n\rangle$ are eigenvectors of N with eigenvalues $n-1$ and $n+1$ respectively, in other words, they are proportional to the eigenvectors $|n-1\rangle$ and $|n+1\rangle$. The proportionality constant is calculated using the norm,

$$|a|n\rangle|^2 = \langle n|a^\dagger a|n\rangle = \langle n|N|n\rangle = n \qquad (3.92)$$

$$\left| a^\dagger |n\rangle \right|^2 = \langle n|aa^\dagger|n\rangle = \langle n|(N+1)|n\rangle = n+1 \qquad (3.93)$$

it is clear that $n \leq 0$, that is, N is defined positive. We can also deduce the formulas,

$$a|n\rangle = \sqrt{n}|n-1\rangle , \quad a^\dagger|n\rangle = \sqrt{n+1}|n+1\rangle \qquad (3.94)$$

Where in the first case n cannot be zero. The operators a and $a^\dagger$ are called destruction and creation operators respectively, namely to the disappearance and appearance of quanta on their action. Notice that in addition $a|0\rangle = 0$.

Using reduction to the absurd, let us imagine that there is a half-eigenvalue q,

$$N|q\rangle = q|q\rangle , \quad n < q < n+1 \qquad (3.95)$$

It is easy to realize that a chain of eigenvectors can be constructed $a^p|q\rangle$ of N with eigenvalues $q - p$,

$$Na|q\rangle = (q-1)a|q\rangle$$
$$Na^2|q\rangle = (q-2)a^2|q\rangle$$
$$\vdots \quad = \quad \vdots$$
$$Na^p|q\rangle = (q-p)a^p|q\rangle \qquad (3.96)$$

and therefore to $a^n|q\rangle$ is eigenvector of N with eigenvalue $q - n > 0$, applying a to we build the non zero eigenvalue, we find,

$$aa^n|q\rangle = (q-n-1)a^n|q\rangle , \quad q-n-1 < 0 \qquad (3.97)$$

which is inconsistent since as we saw, there could be no such negative eigenvalues. Then n must be a nonnegative integer.

In short, we have proved that the Hamiltonian's eigenstates are of the form,

$$H|n\rangle = \left(n + \frac{1}{2}\right)\hbar\omega|n\rangle \qquad (3.98)$$

Which can be generated departing from the ground or zero state as,

$$|n\rangle = \frac{1}{\sqrt{n!}}\left(a^\dagger\right)^n|0\rangle \qquad (3.99)$$

If we want to recover the analytical solution, we can do it by,

$$a|0\rangle = 0 \quad \rightarrow \quad \left(u + \frac{d}{du}\right)|0\rangle = 0 \quad \rightarrow \quad |0\rangle = \sqrt{\frac{\alpha}{\sqrt{\pi}}}e^{-\frac{1}{2}u^2} \quad (3.100)$$

and applying the creation operator is,

$$|n\rangle = \frac{1}{\sqrt{n!}}\frac{1}{2^{\frac{n}{2}}}\left(u - \frac{d}{du}\right)^n|0\rangle$$

$$= \sqrt{\frac{\alpha}{\sqrt{\pi}2^n n!}}H_n(\alpha x)e^{-\frac{\alpha^2}{2}x^2}. \quad (3.101)$$

CHAPTER 4

QUANTUM AXIOMATICS

4.1 INTRODUCTION

Although the development of quantum theory occurred during the first quarter of the twentieth century, which began in 1900 with the discovery of Max Planck quantos, it was not until the 30's when mathematical principles of modern quantum mechanics were established. These principles are fundamentally related to the works of Born, Dirac and John von Neumann.

The axiomatization given by von Neumann lasted for more than half a century in most of the classical books of quantum mechanics. However, in the 1990's the majority of physicists began to adopt a critical attitude towards this form of postulation, especially regarding the intuitive development and the treatment of the problem of collapse of the wave function.

However, it should be noted that the experimental conclusions reached with any set of axioms that can be chosen are, and in fact must be, exactly the same. In fact for most physicists, who do not work on the foundations of quantum mechanics, this is a transparent question.

4.2 THE POSTULATES

4.2.1 Postulate I

Each quantum system will be studied over a complex Hilbert space $\mathcal{H}$, separable and rigged.

Apart from topological considerations, a separable Hilbert space is a vector space $\mathcal{V}$ where a scalar product can be defined, and has a finite linear or infinite numerable basis.

The concept of rigged space is related to topological characteristics that link the square-integrable aspects to those of distribution in functional analysis. This allows us to include a series of generalized functions due to Dirac.

4.2.1.1 Dirac Notation

According to this notation, the vectors of $\mathcal{H}$ are called vectors or kets and are denoted by $|\psi\rangle$. We defined also a dual space $\tilde{\mathcal{H}}$, isomorphic to the above, in which each vector of $\mathcal{H}$ is treated as a linear functional Φ of $\tilde{\mathcal{H}}$ in such a way that each element of $\mathcal{H}$ is associated with a scalar. These vectors are called bra vectors, and are denoted as $\langle\Phi|$,

$$\Phi(\psi) \equiv \langle\Phi|\psi\rangle \in \mathbb{C}. \tag{4.1}$$

In this definition it is assumed that the scalar product defined in each Hilbert space will be used.

Usually in discrete spaces the vectors ket are column vectors and the vectors bra are row vectors. In this sense, it can be seen that, just as a ket by a bra is a c-number or simply a complex number, a product of vector ket with bra will be a q-number or operator.

In each Hilbert space we can define an orthonormal basis $|i\rangle$ such that $\langle i|j\rangle = \delta_{ij}$, using the Kronecker delta or Dirac delta depending if it is the discrete or continuous case. With this basis we can define a relation of completeness or closure of the Hilbert space using the decomposition of the identity,

$$\mathbb{I} = \sum_k |k\rangle\langle k| \tag{4.2}$$

So that any arbitrary vector $|v\rangle$ of the space can be expanded on that basis,

$$|v\rangle = \left(\sum_k |k\rangle\langle k|\right)|v\rangle = \sum_k |k\rangle\langle k|v\rangle = \sum_k v_k|k\rangle \tag{4.3}$$

We can define linear operator on the vectors of Hilbert space, which fulfill,

$$\hat{A}\left(\sum_k a_k|v_k\rangle\right) = \left(\sum_k a_k\hat{A}|v_k\rangle\right) \tag{4.4}$$

These operators have matrix elements given by,

$$A_{ij} = \langle i|\hat{A}|j\rangle \in \mathbb{C}. \tag{4.5}$$

The trace of a linear operator on a given basis is,

$$Tr(A) = \sum_i \langle i|\hat{A}|i\rangle. \tag{4.6}$$

We say that $|v\rangle$ is eigenvector of $\hat{A}$ with eigenvalue v if the equation $\hat{A}||v\rangle = v|v\rangle$ holds. Using algebra we find that the characteristic polynomial is $P(v) = |\hat{A} - vI| = 0$, which does not depend on the chosen representation. The operator will have a diagonal representation in its orthonormal eigenvectors, if any,

$$\hat{A} = \sum_k \lambda_k|k\rangle\langle k| \tag{4.7}$$

The adjunct operator $\hat{A}^\dagger$ is defined as,

$$\langle v|(\hat{A}|w\rangle) = (\langle v|\hat{A})|w\rangle \tag{4.8}$$

An operator is Hermitian or self-joined when obeys the previous equation. An operator $\hat{U}$ is unitary if it holds $\hat{U}\dagger\hat{U} = \mathbb{I}$. These operators are isometric, that is, they preserve the scalar products and are therefore indicated to describe the reversible evolutions.

The unit and self-join operators are part of a larger set, the normal operators, which fulfill $\hat{N}^\dagger\hat{N} = \hat{N}\hat{N}^\dagger$. Normal operators have the peculiarity it is always possible to find a basis of orthonormal eigenvectors, that is, they will be diagonal in that basis.

4.2.2 Spectral Decomposition Theorem

Any normal operator $\hat{N}$ on a Hilbert space $\mathcal{H}$ is diagonal with respect to some orthonormal basis of $\mathcal{H}$.

To demonstrate this, we will prove that any normal operator has a diagonal representation. If we assume that $\hat{P}$ is the projector on the subspace of one of the eigenvalues of $\hat{N}$ and $\hat{Q}$ is complementary projector, such that $\hat{P} + \hat{Q} = \mathbb{I}$. Therefore, we can put,

$$\hat{N} = (\hat{P}+\hat{Q})\hat{N}(\hat{P}+\hat{Q}) = \hat{P}\hat{N}\hat{P}+\hat{P}\hat{N}\hat{Q}+\hat{Q}\hat{N}\hat{P}+\hat{Q}\hat{N}\hat{Q} = \hat{P}\hat{N}\hat{P}+\hat{Q}\hat{N}\hat{Q}. \tag{4.9}$$

We used the fact that $\hat{Q}\hat{N}\hat{P}$ is null, since $\hat{P}$ acting on a vector projects over the subspace of eigenvalues of $\hat{N}$, and $\hat{Q}$ is the orthogonal projector, so the product of the resulting projected vectors will be zero. For annulment of $\hat{P}\hat{N}\hat{Q}$ can be proof similarly, acting on a dual vector on the left, $\hat{P}$ and $\hat{N}$ will leave the dual vector in the orthogonal subspace $\hat{Q}$. Finally, if we prove that $\hat{Q}\hat{N}\hat{Q}$ is normal, as it is applied to the restricted subspace of dimension $n - 1$, it must be admitted that it is diagonal by the principle of induction. To prove it, we use the fact that $\hat{N}$ is normal, that $\hat{Q}^2 = 0$ and the identity $\hat{Q}\hat{N} = \hat{Q}\hat{N}(\hat{P} + \hat{Q}) = \hat{Q}\hat{N}\hat{Q}$ and we have $\hat{Q}\hat{N}\hat{Q}\hat{Q}\hat{N}^{\dagger}\hat{Q} = \hat{Q}\hat{N}\hat{Q}\hat{N}^{\dagger}\hat{Q} = \hat{Q}\hat{N}\hat{N}^{\dagger}\hat{Q} = \hat{Q}\hat{N}^{\dagger}\hat{N}\hat{Q} = \hat{Q}\hat{N}^{\dagger}\hat{Q}\hat{N}\hat{Q} = \hat{Q}\hat{N}^{\dagger}\hat{Q}\hat{Q}\hat{N}\hat{Q}$, then $\hat{Q}\hat{N}\hat{Q}$ is normal. It is obvious that $\hat{P}\hat{N}\hat{P}$ is already diagonal on any base that contains the vector belonging to $\hat{P}$. It follows that any normal, or self-joined, unitary operator can always be decomposed in terms of its eigenvectors or the orthogonal projectors on them, in the form,

$$\hat{O} = \sum_k \lambda_k |k\rangle\langle k| = \sum_k \lambda_k P_k, \tag{4.10}$$

being $|k\rangle$ the basis of eigenvectors with eigenvalues λ_k and $\hat{P}_k$ the projectors onto subspaces of eigenvalue λ_k.

4.2.3 Postulate II

A state is represented by an operator ρ, called density matrix, that is non-negative, self-join, and with trace equals to one. The observables will also be self-joined operators, and their eigenvalues will be the possible values of the physical magnitudes.

4.2.3.1   Expression of States: Matrix Density and Vectors

The fact that the state operator is self-adjoint ensure us that it can be decomposed into a spectral representation of the form,

$$\hat{\rho} = \sum_k p_k |\psi_k\rangle\langle\psi_k|, \tag{4.11}$$

in terms of their eigenvalues and orthonormal eigenvectors, where we have assumed a discrete case for convenience.

Besides this conditions it must be fulfilled that,

$$0 \le p_k \le 1, \qquad \sum_k p_k = 1. \tag{4.12}$$

Within this mixture of states there is the particular case of pure states,

$$\rho = |\psi\rangle\langle\psi|. \tag{4.13}$$

In this case, the state of the system is often associated not with its density matrix but with the so-called state vector $|\psi\rangle$, which being a unit ray in Hilbert space will have a phase ambiguity,

$$|\psi\rangle \sim e^{i\alpha}|\psi\rangle. \tag{4.14}$$

However, the state operator is independent of this arbitrary phase. It can be proved that these pure states cannot be represented by combinations of other states.

It should be noted that the representation of a non-pure state as a combination of pure states is not unique and depends on the basis we use, which not necessarily will be the basis of orthonormal eigenvectors.

4.2.3.2   Complete Set of Commuting Operators or C.S.C.O.

Theorem

If A and B are two self-joined operators that have a complete set of eigenvectors ,and if the operators commute, i.e., $[\hat{A}, \hat{B}] = \hat{A}\hat{B} - \hat{B}\hat{A} = 0$, there is a complete set of eigenvectors common to both.

To demonstrate, let two sets of eigenvectors be,

$$\hat{A}|a_n\rangle = a_n|a_n\rangle \tag{4.15}$$

$$\hat{B}|b_m\rangle = b_m|b_m\rangle. \tag{4.16}$$

We can expand $|a_n\rangle$ in the base of eigenvectors of $\hat{B}$,

$$|a_n\rangle = \sum_m c_m|b_m\rangle \tag{4.17}$$

then,

$$\left(\hat{A} - a_n\right)|a_n\rangle = 0 = \sum_m \left(\hat{A} - a_n\right)c_m|b_m\rangle. \tag{4.18}$$

If we operate with $\hat{B}$ on the left and using the commutativity in each term of the summation we obtain,

$$\hat{B}\left(\hat{A} - a_n\right)|b_m\rangle = b_m\left(\hat{A} - a_n\right)|b_m\rangle. \tag{4.19}$$

That is, $(\hat{A} - a_n)|b_m\rangle$ is an eigenvector of $\hat{B}$ with eigenvalue b_m, and therefore all summation terms will be orthogonal and will have to be annulled separately, which means that $|b_m\rangle$ will also be eigenvalue of $\hat{A}$.

In general the group of these operators in which there is no degeneracy, i.e., for each eigenvalue there exists a single eigenvector, is called a complete set of commuting operators.

Theorem
Any operator that commutes with a complete set of switching operators will be a function of the operators of that set. Let be the set $(\hat{A}, \hat{B}, ...)$ of operators with eigenvectors $|a_n, b_m...\rangle$,

$$\hat{A}|a_n, b_m, ...\rangle = a_n|a_n, b_m, ...\rangle \tag{4.20}$$

$$\hat{B}|a_n, b_m, ...\rangle = b_m|a_n, b_m, ...\rangle, \tag{4.21}$$

and let $\hat{F}$ an operator that commutes with all. By the previous theorem a complete set would be the extended $(\hat{A}, \hat{B}, ..., \hat{F})$, but the vectors $|a_n, b_m...\rangle$ are the only eigenvectors of the C.S.C.O. in that Hilbert space, then they must also be eigenvectors of the extended,

$$\hat{F}|a_n, b_m, ...\rangle = f_{nm..}|a_n, b_m, ...\rangle \tag{4.22}$$

And therefore there must be a function such that:

$$f_{nm..} = f(a_n, b_m, ...). \tag{4.23}$$

4.2.4   Postulate III

The mean value of a dynamic variable O (which can represent an observable) in the set of equally prepared states represented by the operator ρ is

$$\langle \hat{O} \rangle_\rho = Tr(\rho \hat{O}), \tag{4.24}$$

and will also give the probability distribution in the Hilbert space basis considered.

4.2.4.1   Mean Values

Let's look at the general expression that has the mean value of an observable by means of a certain orthonormal basis,

$$\langle \hat{O} \rangle_\rho = Tr(\rho \hat{O}) = \sum_k \langle \psi_k | \rho \hat{O} | \psi_k \rangle = \sum_{i,k} \langle \psi_k | \psi_i \rangle p_i \langle \psi_i | \hat{O} | \psi_k \rangle$$
$$= \sum_k p_k \langle \psi_k | \hat{O} | \psi_k \rangle, \tag{4.25}$$

and if the state is a pure state, represented by the vector $|\psi\rangle$ the mean value would be reduced to,

$$\langle \hat{O} \rangle = \langle \psi | \hat{O} | \psi \rangle. \tag{4.26}$$

It can be said in general that the average value of the observable in any state is the statistical average of its mean values in each pure state and whose weights are precisely the coefficients of the decomposition of the state in that basis.

4.2.4.2 Probability

Let us assume that we want to calculate the probability $P(a_n)$ that a measurement $\hat{A}$ yields a value a_k. Let us write the density operator of the form,

$$\hat{\rho} = \sum_k p_k |\psi_k\rangle\langle\psi_k|. \tag{4.27}$$

The probability to obtain a_n is found using,

$$P(a_n) = Tr\left(\hat{\rho}I(a_n)\right), \qquad I(a_n) = |a_n\rangle\langle a_n|. \tag{4.28}$$

If $|m\rangle$ is a complete basis, i.e., $\sum_m |m\rangle\langle m| = 1$, then,

$$Tr\left(\hat{\rho}I(a_n)\right) = \sum_m \langle m|\hat{\rho}|a_n\rangle\langle a_n|m\rangle \tag{4.29}$$

we can expand, obtaining,

$$\sum_m \langle m|\hat{\rho}|a_n\rangle\langle a_n|m\rangle = \sum_m \sum_k p_k \langle m|\psi_k\rangle\langle\psi_k|a_n\rangle\langle a_n|m\rangle \tag{4.30}$$

$$= \sum_m \sum_k p_k \langle a_n|m\rangle\langle m|\psi_k\rangle\langle\psi_k|a_n\rangle \tag{4.31}$$

$$= \sum_k p_k \langle a_n|\psi_k\rangle\langle\psi_k|a_n\rangle \tag{4.32}$$

$$= \sum_k p_k |\langle a_n|\psi_k\rangle|^2 \tag{4.33}$$

It should be noted that the probability would have absolute certainty in the case that it is in an eigenstate of the observable and a measure of the corresponding eigenvalue is performed, and will be zero in the case that we are in pure orthonormal state to this.

CHAPTER 5

QUANTUM MEASUREMENTS

5.1 INTRODUCTION

The traditional response to the problem of measurement is based on the so-called "collapse hypothesis," first formulated by Werner Heisenberg in an article in 1927, in terms of "wave packet reduction." According to this hypothesis, quantum systems develop two types of temporal evolution, a deterministic and unitary evolution, when they are not observed, and an indeterministic and non-unitary transition or "collapse" of the wave function, that occurs when measurement is performed.

Such a transition would lead the system from its original superposition state to another state, in which the system and the measuring apparatus acquire definite properties.

Nowadays, the problem of measurement is approached from the theory of environment-induced decoherence. This theoretical program, based on the study of the effects of the interaction between a quantum system, is considered as an open system and its environment. According to this theory, the system-environment interaction leads to a decoherence process that selects the "observables" properties of the system.

In turn, even the measurement problem has been reformulated, expanding its scope. As Schlosshauer points out, the conceptual difficulties of quantum measurement can now be concentrated around two cases:

- The definite outcome, which consists of the traditional question about quantum measurement.

- The problem of the preferred basis: since there is a theoretical ambiguity in the definition of the observable measured due to the mathematical possibility of a base change in the expression of the system state.

5.2   THE QUANTUM MEASUREMENT

In the traditional formulation due to John von Neumann, a quantum measurement is conceived as an interaction between the system to be measured S and a system M, the measuring apparatus. The S system, represented in the Hilbert space $\mathcal{H}_S$, has an observable $\hat{A} = \sum_k a_k |a_k\rangle\langle a_k|$, where a_k are eigenvalues of $\hat{A}$ and $\{|a_k\rangle\}$ is a basis of $\mathcal{H}_S$. The apparatus M, has an observable $\hat{P} = \sum_k p_k |p_k\rangle\langle p_k|$, where the p_k are eigenvalues of $\hat{P}$ and $\{|pi\rangle\}$ is a base on $\mathcal{H}_S$. In order to fulfill the role of the apparatus, $\hat{P}$ must have different and macroscopically distinguishable eigenvalues. Suppose that, prior to the interaction, S is in a superposition of the eigenvectors of the observable $\hat{A}$, and M is in a base state $|p_0\rangle$ that can be defined as initial,

$$|\psi_0\rangle = \sum_k c_k |a_k\rangle \otimes |p_0\rangle. \qquad (5.1)$$

The composite system will evolve, according to the Schrödinger equation, to a new state $|\psi_f\rangle$, where the eigenvectors of the observable $\hat{A}$ in S are correlated with the eigenvectors of the observable $\hat{P}$ in the apparatus M,

$$|\psi_0\rangle = \sum_k c_k |a_k\rangle \otimes |p_0\rangle \rightarrow |\psi_f\rangle = \sum_k c_k |a_k\rangle \otimes |p_k\rangle. \qquad (5.2)$$

However, despite the perfect correlation, the difficulty lies in the nature of the final state $|\psi_f\rangle$, which is an overlapping of states $|a_k\rangle \otimes |p_k\rangle$. Therefore, without an additional interpretive assumption, there are no arguments to account for the fact that the pointer $\hat{P}$ of the apparatus acquires one of its eigenvalues p_k in a definite way. Historically, the collapse hypothesis was the first response to the problem of a defined outcome. According to this hypothesis, at the instant of measurement, the composite system $S + M$

have an indeterministic and non-unitary transition that converts the state $|\psi_f\rangle$ into one of the states of the overlap, say $|\phi_k\rangle$

$$|\psi_f\rangle = \sum_k c_k |a_k\rangle \otimes |p_k\rangle = \sum_k |\phi_k\rangle \to |\phi_k\rangle |a_k\rangle \otimes |p_k\rangle. \qquad (5.3)$$

Since $|\phi_k\rangle$ is no longer an overlap, it is inferred that the observable $\hat{A}$ acquires the value a_k and the observable $\hat{P}$ acquires the value p_k. Since collapse is an indeterministic process, the probability that $|\psi_f\rangle$ "collapse" in each state $|\phi_k\rangle$ is $|c_k|^2$. This means that if many particular measurement is made on the same system under the same initial condition, each state $|\phi_k\rangle$ is obtained with a frequency that will approximate $|c_k|^2$ as the number of measurements increases. Thus,

$$\rho_m = \sum_k |c_k|^2 |a_k\rangle \otimes |p_k\rangle \angle a_k| \otimes \angle p_k|. \qquad (5.4)$$

Always according to the collapse hypothesis, the density operator ρ_m represents a mixture states interpreted in terms of ignorance, that is, it can be interpreted as expressing the system as actually being in some of states $|a_k\rangle \otimes |p_k\rangle$, but the observer does not have information about what particular state, the only known information is through the probability $|c_k|^2$ corresponding to each one of the states. In this way, if it does happen $|a_i\angle \otimes |p_i\langle$, we could say that the observable $\hat{P}$ of the apparatus, acquires a defined value p_i, and this would indicate that the system S is in the state $|a_i\rangle$.

The main objection directed against this interpretation of quantum measurement refers to the ad hoc character of the hypothesis: collapse would be a physical process which, however, does not account for the theory. In this way the type of interaction that would give rise to this process, as well as the precise moment in which it would occur, is undefined. Therefore, the collapse hypothesis is not adopted in its original form in the discussions about the foundations of quantum mechanics.

Many interpretations have been specifically designed to provide a conceptually appropriate solution to the problem. The theory of decoherence, however, is presented as an approach that, from the theoretical physics itself, solves the difficulties of quantum measurement.

5.2.1   The Theory of Decoherence

Consider how decoherence acts in quantum measurement (Paz & Zurek, 2002). As we pointed out in the previous section, after the interaction between the system S and the apparatus M, the composite system $S + M$ is in a state of overlap:

$$|\psi_f\rangle = \sum_k c_k |a_k\rangle \otimes |p_k\rangle \tag{5.5}$$

Let us assume that the observable $\hat{A}$ has only two self-states, $|a_1\rangle$ and $|a_2\rangle$, such that the state $|\psi_f\rangle$,

$$|\psi_f\rangle = c_1 |a_1\rangle \otimes |p_1\rangle + c_2 |a_2\rangle \otimes |p_2\rangle \tag{5.6}$$

According to decoherence theory, this system is not isolated but in interaction with an environment E with a very large number of degrees of freedom. Suppose E is made up of a huge number N of spin particles that do not interact with each other. Therefore, at an initial time $t = 0$, the state of the composite system $C = S + M + E$ results in,

$$|\psi_C(t=0)\rangle = (c_1 |a_1\rangle \otimes |p_1\rangle + c_2 |a_2\rangle \otimes |p_2\rangle) \otimes \prod_{i=1}^{N} (d_k | \uparrow\rangle_k + f_k | \downarrow\rangle_k) \tag{5.7}$$

where $| \uparrow\rangle_k$ and $| \downarrow\rangle_k$ are the self-states of the k component of the spin corresponding to the particle k. For simplicity, it is considered that the Hamiltonian of the apparatus and the environment are zero and in addition, the interaction between S and M has ceased. Therefore, the evolution of the composite system C will be governed by the interaction Hamiltonian $\hat{H}_I$. If it is assumed that

$$\hat{H}_I = (|a_1\rangle \otimes |p_1\rangle\langle p_1| \otimes \langle a_1| + |a_2\rangle \otimes |p_2\rangle\langle p_2| \otimes \langle a_2|)$$
$$\otimes \sum_k g_k (| \uparrow\rangle\langle\uparrow | + | \downarrow\rangle\langle\downarrow |)_k \tag{5.8}$$

Under the influence of $\hat{H}_I$ the initial state $|\psi_C(t = 0)\rangle$ will evolve as,

$$|\psi_C(t = 0)\rangle = c_1 |a_1\rangle \otimes |p_1\rangle \otimes |\epsilon_1(t)\rangle + c_2 |a_2\rangle \otimes |p_2\rangle \otimes |\epsilon_2(t)\rangle \tag{5.9}$$

where,

$$|\epsilon_1(t)\rangle = \prod_k \left(d_k e^{ig_k t}|\uparrow\rangle_k + d_k e^{-ig_k t}|\downarrow\rangle_k\right) \tag{5.10}$$

$$|\epsilon_2(t)\rangle = \prod_k \left(d_k e^{-ig_k t}|\uparrow\rangle_k + d_k e^{ig_k t}|\downarrow\rangle_k\right) \tag{5.11}$$

In this way, the state of the $S + M$ system is "entangled" with the states of the environment E. If we recall that the density operator corresponding to a pure state $|j\rangle$ is defined as $\rho = |j\rangle\langle j|$, the density operator $\rho_C(t)$ corresponding to this interlaced state will be,

$$\rho_C(t) = |\psi_C(t)\rangle\langle\psi_C(t)| \tag{5.12}$$

$$= |c_1|^2 |a_1\rangle \otimes |p_1\rangle \otimes |\epsilon_1(t)\rangle\langle a_1| \otimes \langle p_1| \otimes \langle \epsilon_1(t)|$$

$$+ c_1^* c_2 |a_2\rangle \otimes |p_2\rangle \otimes |\epsilon_2(t)\rangle\langle a_1| \otimes \langle p_1| \otimes \langle \epsilon_1(t)|$$

$$+ c_1 c_2^* |a_1\rangle \otimes |p_1\rangle \otimes |\epsilon_1(t)\rangle\langle a_2| \otimes \langle p_2| \otimes \langle \epsilon_2(t)|$$

$$+ |c_2|^2 |a_2\rangle \otimes |p_2\rangle \otimes |\epsilon_2(t)\rangle\langle a_2| \otimes \langle p_2| \otimes \langle \epsilon_2(t)|$$

$$\tag{5.13}$$

where the crossed terms represents the quantum correlations that prevents the interpretation of $\rho_C(t)$ in classical terms.

At any time t the description of the system $S+M$ is given by the reduced density operator $\rho_r(t) = Tr_{(}E)[\rho_C(t)] = S\langle e_i(t)|\rho_C(t)|e_i(t)\rangle$, which is obtained eliminating the degrees of freedom of the environment by the partial trace mathematical operation $Tr_{(}E)$ (see Ballentine, 1998). Thus,

$$\rho_r(t) = |c_1|^2 |a_1\rangle \otimes |p_1\rangle\langle a_1| \otimes \langle p_1| + \xi(t)c_1 c_2^* |a_1\rangle \otimes |p_1\rangle\langle a_2| \otimes \langle p_2| +$$

$$+\xi(t)^* c_1^* c_2 |a_2\rangle \otimes |p_2\rangle\langle a_1| \otimes \langle p_1| + |c_2|^2 a_2\rangle \otimes |p_2\rangle\langle a_2| \otimes \langle p_2|$$

$$\tag{5.14}$$

where the factor $\xi(t)$,

$$\xi(t) = \langle\epsilon_1(t)|\epsilon_2(t)\rangle = \prod_k \left[\cos(2g_k t) + i\left(|d_k|^2 - |f_k|^2\right)\sin(2g_k t)\right],$$

$$\tag{5.15}$$

determines at each instant the value of the terms outside the diagonal. For N large enough, as time passes the states rapidly tend towards orthogonality so that, in a very short interval $\xi(t) \to 0$.

This means that, for an extremely small "decoherence time," the reduced density operator $\rho_r(t)$ converges to,

$$
\begin{aligned}
\rho_r(t) = & |c_1|^2 |a_1\rangle \otimes |p_1\rangle\langle a_1| \otimes \langle p_1| \\
& + |c_2|^2 a_2\rangle \otimes |p_2\rangle\langle a_2| \otimes \langle p_2|
\end{aligned}
\tag{5.16}
$$

where the terms outside the diagonal disappeared. Accordingly, $\rho_r(t)$ denotes a mixture state that contains only the terms corresponding to the classical correlations and, therefore, can be interpreted in terms of ignorance. The compound system $S + M$ is found in some of the states $|a_1\rangle \otimes |p_1\rangle$ or $|a_2\rangle|p_2\rangle$, and the probabilities $|c_1|^2$ and $|c_2|^2$ measure our lack of knowledge about the defined state of the system. As a consequence, the environment-induced decoherence would lead to the same state ρ_m introduced by the collapse hypothesis; but without assuming an additional physical process to the unitary evolution described by the Schrödinger equation.

5.2.2 The Problem of the Privileged Base

The difficulty of a base would lie in the theoretical ambiguity in the definition of the measured observable, due to the mathematical possibility of expressing the state of the system in different bases of a Hilbert space. Let us look at the problem as formulated by Zurek (1981).

Recalling von Neumann's formulation of the problem of quantum measurement that states,

$$
|\psi_0\rangle = \sum_k c_k |a_k\rangle \otimes |p_0\rangle \rightarrow |\psi_f\rangle = \sum_k c_k |a_k\rangle \otimes |p_k\rangle.
\tag{5.17}
$$

Through interaction, the states $|a_k\rangle$ of the system S are perfectly correlated with the states $|p_k\rangle$ of the measuring apparatus M. According to Zurek (1981), precisely because of this that we claim the observable $\hat{A} = Sa_i|a_i\rangle\langle a_i|$ is the product of the measurement of $\hat{A}$. However, Zurek (1981) assures, that in certain circumstances the formalism of quantum mechanics allows us to express the final state $|f >$ in another basis of the Hilbert space of the composite system, say, $\{|a_i'\angle \otimes |p_i'\rangle\}$,

$$|\psi_f\rangle = \sum_k c_k |a_k\rangle \otimes |p_k\rangle \to \sum_k c_k' |a_k'\rangle \otimes |p_k'\rangle. \qquad (5.18)$$

The possibility of expressing the same state in terms of self-states of another observable would thus introduce a fundamental ambiguity: without some additional element we would not be able to determine the privileged base of the apparatus, that is, in what mixture the wave function collapses properly. Therefore, we would not know which observable was measured and also which are the states of the measurement apparatus, that is, what set of outcomes will be obtained in the measurement.

It is not difficult to imagine the answer to the problem thus posed in the framework of the decoherence program. It is precisely decoherence that results from the interaction with the environment the key that would solve the ambiguity in the choice of the privileged base. In fact, as we saw in the previous section, when the composite system $S + M$ interacts with the environment E, in extremely short times the quantum correlations disappear and the composite system acquires a final state.

In this way, we could say that on the S system the observable $\hat{A}$ has been measured by $\hat{P}$. In other words, the interaction with the environment would be the additional element that would determine what mix the wave function seems to have collapsed and, with this, would univocally identify the observable that has been measured on S.

In the case of the privileged base problem, the solution proposed by Zurek (1981) has achieved much greater consensus. For example, according to Schlosshauer (2004), the clear merit of the environment-induced decoherence approach is that the privileged base is not chosen in an ad hoc way to make our observations definite. Selection is based on physical and independent bases of the observer. The value of decoherence lies in its ability to select a special basis. It is precisely for this reason that decoherence has been invoked to solve the particular difficulties of certain interpretations of quantum mechanics. For example, Schlosshauer argues that "it is reasonable to anticipate that decoherence, immersed in some additional interpretive structure, can lead to a complete and consistent description of the classical world from quantum-mechanical principles" (2004, P. 1287).

It is clear that the decoherence identifies a privileged base through the resource of the interaction with the environment. It is also true that with this the decoherence theory could provide an important element to certain realistic interpretations, such as that of many worlds, whose weak is precisely the lack of a criterion to define the privileged base.

5.2.3 The Example of the Stern-Gerlach Type Measurement

In his presentation of the problem of the privileged base, Zurek (1981) points to the typical case of the Stern-Gerlach measurement, where the moment P_z in the z direction plays the role of base state of $\hat{P}$, and the state of the system is expressed in the basis of spin S_z in the z direction,

In this articular case the initial state $|S\rangle$ is,

$$|\psi_S\rangle = \frac{1}{\sqrt{2}}(|\uparrow\rangle + |\downarrow\rangle). \tag{5.19}$$

In this case, the process of interaction between S and M can be expressed as,

$$|\psi_0\rangle = \frac{1}{\sqrt{2}}(|\uparrow\rangle + |\downarrow\rangle) \otimes |0\rangle \rightarrow |\psi_f\rangle = \frac{1}{\sqrt{2}}(|\uparrow\rangle \otimes |p_+\rangle + |\downarrow\rangle \otimes |p_-\rangle) \tag{5.20}$$

where the p_+ and p_- comes from, $\hat{P}_z|\pm\rangle = p_\pm|\pm\rangle$.

But the state $|\psi_f\rangle$ can be expressed in another basis, for example, as,

$$\begin{aligned}|\psi_f\rangle &= \frac{1}{\sqrt{2}}(|\uparrow\rangle \otimes |p_+\rangle + |\downarrow\rangle \otimes |p_-\rangle) \\ &= \frac{1}{\sqrt{2}}(|\rightarrow\rangle \otimes |p'_+\rangle + |\leftarrow\rangle \otimes |p'_-\rangle)\end{aligned} \tag{5.21}$$

where $|\rightarrow\rangle$ and $|\leftarrow\rangle$ are the self-vectors of the spin in the direction x and y, $|p'_+\rangle$ and $|p'_- >$ are the eigenvectors of a new $\hat{P}'$.

However, this apparently innocent example hides peculiarities which, when brought to light, reveal the weaknesses of the general argument.

First, the fact that the base change expressed in the particular case of Eq. (5.21) is correct does not imply that the change in generic basis is always possible in a non-trivial way. Using the bi-orthogonal decomposition theorem or Schmidt's theorem we know that the decomposition of $|\psi_f\rangle \in H_S \otimes H_M$ into self-states $|a_k\rangle \otimes |p_k\rangle$, where $\{|a_k >\}$ is a base of H_S and $\{|pi>\}$ is a base of H_M. It is unique in the non-degenerate case, when all the coefficients $|c_k|$ are different from each other. Thus, the base change expressed in equation is only possible degenerate case, that is, when some of the coefficients $|c_k|$ are equal in absolute value.

The case of a Stern-Gerlach measurement in which the initial state of the system S to measure is,

$$|\psi_S\rangle = \frac{1}{\sqrt{2}}(|\uparrow\rangle + |\downarrow\rangle) \tag{5.22}$$

where $c_1 = c_2 = 1/2$. It is in this situation that the alleged problem of the base ambiguity manifest itself. Therefore, the problem of the privileged base is general. It would manifest for some initial states of the S system, as in the example. Secondly, one wonders how important the initial degenerate states of the system are for measuring. In general, the initial state $|\psi_S\rangle$ of S is expressed in the basis defined by the observable S_z as,

$$|\psi_S\rangle = c_1|\uparrow\rangle + c_2|\downarrow\rangle, \tag{5.23}$$

where c_1 and c_2 are any complex numbers satisfying the normalization condition $|c_1|^2 + |c_2|^2 = 1$. It is easy to check, that the space of the pairs $(|c_1|, |c_2|)$ that satisfy the condition $|c_1| = |c_2|$, has a dimension smaller than the space of the pairs $(|c_1|, |c_2|)$ that only fulfill the normalization condition. Therefore, the degenerate initial states of the system are particular cases that form a void space of measurement in the space of all possible initial states $|\psi_S\rangle$.

5.2.3.1 What is a Quantum Measurement?

It is true that a Stern-Gerlach measurement it is often referred as a "spin measurement," but this reference is only an analogy with classical measurement, where the correlation between an observable $\hat{A}$ of the system and the operator $\hat{P}$ of the measuring apparatus allows the value of $\hat{A}$ to be known through the value of $\hat{P}$. Experimental physicist knows that the goal of a quantum measurement is not well known. There are theoretical difficulties entailed by the assignment of a precise value to an observable when the state of the system is not an eigenvector. The purpose of a quantum measurement is to reconstruct the state $|\psi_S\rangle$ in which the S system was before the interaction. To achieved this, many detections will be necessary, so that the frequency with which each eigenvalue p_k of the operator $\hat{P}$ is detected gives the values $|c_k|^2$. Therefore, the measuring apparatus is designed to obtain the values $|c_k|^2$ for any values c_k, that is, for any $|\psi_S\rangle$. In other words, it is precisely the initial state $|\psi_S\rangle = Sc_k|a_k\rangle$ that is in principle unknown and that is intended to reconstruct from the quantum measurement.

Once the specificity of the quantum measurement is understood and the result of Schmidt's theorem is recalled, the basis ambiguity argument loses its original plausibility. In effect, the experimental arrangement is designed to reconstruct the initial state of the S system, whatever the initial state. The fact that for some particular cases of $|\psi_S\rangle$ the non-trivial base change cannot be made implies some ambiguity. The measurement is not designed to measure a particular state of the system S, but must be effective for any $|\psi_S\rangle$. Thus, when the basis ambiguity argument is analyzed in the light of these considerations, it is easy to verify that Zurek constructs the privileged basis problem by ignoring the nature of quantum measurement and assuming that the problem is reduced to choosing a basis for a quantum state. But a measurement involves an experimental arrangement designed to reconstruct, at least partially, the initial state of the system to be measured, and therefore, there is a single basis that retains the necessary correlation for any values c_k. Consequently, the $\hat{P}$ remains uniquely defined.

5.2.3.2   Measurement as a Process

The argument of the ambiguity of the privileged basis is presented as a consequence of a mathematical result, the possibility of expressing a vector of the Hilbert space on any basis. However, this presentation of the problem ignores that a quantum measurement is a process that originates at the initial state of the system and, as a process, must be governed by the dynamics of the postulates of quantum mechanics, that is, the equation of Schrödinger.

Recall that, according to the Von Neumann model, the composite system $S + M$ is initially found in a state $|\psi_0\rangle$. Through the interaction, $|\psi_0\rangle$ evolves to $|\psi_f\rangle$ where the eigenvectors of the observable $\hat{A}$ in the system are correlated with the eigenvectors of the observable $\hat{P}$ in the apparatus

M. According to the Schrödinger equation, Such evolution is determined by a unitary evolution operator $\hat{U}(y)$,

$$|\psi_0\rangle = \sum_i c_i |a_i\rangle \otimes |p_0\rangle \rightarrow |\psi_f\rangle = \hat{U}(t)|\psi_0\rangle = \sum_i c_i |a_i\rangle \otimes |p_i\rangle. \quad (5.24)$$

According to the argument of the base ambiguity, we measure the observable by the $\hat{P}$. However, since $|\psi_f\rangle$ can also be expressed in the base $\{|a_i'\rangle \otimes |P_i'\rangle\}$, we also have measured the observable A' by P'. To analyze the argument from the perspective of measurement as a process, we see how the measurement of A "by $\hat{P}$" would be expressed,

$$|\psi_0 = \sum_k c_k' |a_k'\rangle \otimes |p_0\rangle \rightarrow |\psi_f\rangle = \hat{U}'(t)|\psi_0'\rangle = \sum_k c_k' |a_k'\rangle \otimes |p_k'\rangle. \quad (5.25)$$

where we know that $|\psi_f\rangle = |\psi_f'\rangle$, since it is the same vector expressed in different bases. Thus,

$$\hat{U}(t)|\psi_0\rangle = \hat{U}'(t)|\psi_0'\rangle. \quad (5.26)$$

This reveals that, in general, expressions (5.24) and (5.25) describe different measurements, because although they lead to the same final state, they start from different initial states $|\psi_0\rangle$ and $|\psi_0'\rangle$ that evolve according to different evolution operators $\hat{U}(t)$ and $\hat{U}'(t)$.

Now, the evolution operator is a function of the interaction Hamiltonian $\hat{H}_I$ between S and M, defined by the physical characteristics of the apparatus M,

$$\hat{U}(t) = \hat{U}'(t) = e^{-i\hat{H}_I t}. \quad (5.27)$$

The Hamiltonian introduces the correlation between the eigenstates of the observable $\hat{A}$ and the eigenstates of $\hat{P}$ of the apparatus. Because of this, $\hat{H}_I$ must have a very precise form, which depends on $\hat{A}$ and $\hat{P}$ [121],

$$\hat{H}_I = \frac{\lambda \hbar}{\Delta t} \left(\hat{A} \otimes \hat{R}_P \right). \quad (5.28)$$

where λ is a constant, Δt is the duration of the interaction and $\hat{R}_P$ is the canonically conjugate observable of $\hat{P}$, with $[\hat{P}, \hat{R}_P] = i\hbar$. If $\hat{U}(t)$ and $\hat{U}'(t)$ express the same evolution operator and correlate different bases, then the interaction Hamiltonian $\hat{H}_I$ should also correlate the eigenstates of the observable $\hat{A}'$ and the those of $\hat{P}'$ therefore, should also be able to be expressed as,

$$\hat{H}_I = \frac{\lambda \hbar}{\Delta t} \left(\hat{A}' \otimes \hat{R}_{P'} \right).$$

(5.29)

where, by hypothesis, $\hat{A} \neq \hat{A}'$ and $\hat{P} \neq \hat{P}'$. Therefore, it should be observed that

$$\hat{A}' \otimes \hat{R}_{P'} = \hat{A} \otimes \hat{R}_P.$$

(5.30)

In this way, we arrive to a contradiction. In short, this argument shows that the observable that is measured is not determined by the basis on which the final state of the measurement is expressed, but by the process that, starting from the initial state, establishes the correlations. Regardless of the initial state, this process is unequivocally defined by the evolution operator $\hat{U}(t) = e^{-i\hat{H}_I t}$ which is fixed by the construction of the measuring apparatus. In other words, it is the measuring apparatus which sets the privileged basis to the extent that its $\hat{H}_I$ correlates $\hat{A}$ and $\hat{P}$, and not any other pair $\hat{A}'$ and $\hat{P}'$.

CHAPTER 6

PATH INTEGRAL FORMULATION OF QUANTUM MECHANICS

To introduce the formalism of the functional integral in quantum mechanics, we start from the definition of the temporal evolution operator. In quantum mechanics, time evolution is determined by the time-dependent Schrödinger equation,

$$i\hbar\frac{\partial}{\partial t}|\psi(t)\rangle = \hat{H}|\psi(t)\rangle. \tag{6.1}$$

where $\hat{H}$ is the Hamiltonian of the system and $|\psi_{(}t)\rangle$ is the ket which describes a state at time t. Introducing the wave function $\psi(x,t) = \langle x|\psi(x,t)\rangle$ in the coordinates representation, the equation of Schrödinger takes the form,

$$i\hbar\frac{\partial}{\partial t}\psi(x,t) = -\frac{\hbar^2}{2m}\frac{\partial^2}{\partial x^2}\psi(x,t) + V(x,t)\psi(x,t) \tag{6.2}$$

where we have assumed that Hamiltonian is of the form $\hat{H} = \frac{p^2}{2m} + \hat{V}(x,t)$. Solving this equation means finding the time evolution operator $\hat{U}(t_2,t_1)$ that propagates the state of the system from t_1 to t_2,

$$|\psi(t_2)\rangle = U(t_2,t_1)|\psi(t_1)\rangle \tag{6.3}$$

In the case that $\hat{H}$ does not explicitly depend on time is given by,

$$U(t_2, t_1) = U(t_2 - T_1) = e^{-\frac{i}{\hbar}(t_2 - t_1)H} \tag{6.4}$$

In the most general case, $\hat{U}(t_2, t_1)$ can be determined using the Dyson series and written in a compact form,

$$U(t_2, t_1) = Te^{-\frac{i}{\hbar}\int_{t_1}^{t_2} d\tau H(\tau)} \tag{6.5}$$

where with T indicates the temporarily ordered product of operators. This is such that, retains the order the operators with respect to time so are arranged in descending order from right to left, for example, we have,

$$T(A(t_1)B(t_2)) = \begin{cases} A(t_1)B(t_2) & \text{if } t_1 > t_2 \\ B(t_2)A(t_1) & \text{if } t_2 > t_1 \end{cases} \tag{6.6}$$

In the coordinates representation of the $\hat{U}(t_2, t_1)$ is determined completely if you know its matrix elements

$$\langle x_2 | U(t_2, t_1) | x_1 \rangle \equiv U(t_2, x_2; t_1, x_1) \tag{6.7}$$

and the time evolution of the wave function is obtained by,

$$\psi(x_2, t_2) = \int dx_1 U(t_2, x_2; t_1, x_1) \psi(x_1, t_1) \tag{6.8}$$

On the other hand, in what is called the Heisenberg scheme, operators are evolving in time, while states do not change. Using the relationship given by timer evolution operator,

$$|\psi(t)\rangle = U(t, 0)|\psi\rangle_H \tag{6.9}$$

And by exploiting the unitarity of $\hat{U}(t)$ is obtained that,

$$_H\langle x_2, t_2 | x_1, t_1 \rangle_H = \langle x_2 | U(t_2, 0) U^\dagger(t_1, 0) | x_1 \rangle \tag{6.10}$$

And as it is the law of composition

$$U(t_2, 0) = U(t_2, t_1) U(t_1, 0) \tag{6.11}$$

we obtain,

$$_H\langle x_2, t_2 | x_1, t_1 \rangle_H = U(t_2, x_2; t_1, x_1) \quad (6.12)$$

This means that the time evolution operator matrix elements are the transitions amplitudes between the eigenstates of the operator position in the representation of Heisenberg.

6.1 CALCULATING THE TRANSITION AMPLITUDE

To obtain $\hat{U}\langle t_f, x_f; t_i, x_i \rangle = \langle x_f, t_f | x_i, t_i \rangle$ with $t_f > t_i$ in an alternate way. Consider a time-independent Hamiltonian and let us split an interval (t_i, t_f) in N pieces of equal length given by $\epsilon = \frac{t_f - t_i}{N}$ such that $t_f = t_i + n\epsilon$ with $n = 1, 2, .., N - 1$.

Introducing a complete set eigenstates for each intermediate time,

$$U(t_f, x_f; t_i, x_i) = \langle x_f, t_f | x_i, t_i \rangle$$
$$= \lim_{\substack{\epsilon \to 0 \\ N \to \infty}} \int dx_1 \cdots dx_{N-1} \langle x_f, t_f | x_{N-1}, t_{N-1} \rangle (6.13)$$
$$\langle x_{N-1}, t_{N-1} | x_{N-2}, t_{N-2} \rangle \cdots$$
$$\cdots \langle x_2, t_2 | x_1, t_1 \rangle \langle x_1, t_1 | x_i, t_i \rangle \quad (6.14)$$

The transition amplitude $\langle x_f, t_f | x_i, t_i \rangle$ in terms of a sequence of intermediate amplitudes. A general expression is,

$$\langle x_n, t_n | x_{n-1}, t_{n-1} \rangle = \langle x_n | e^{-\frac{i}{\hbar} t_n H} e^{-\frac{i}{\hbar} t_{n-1} H} | x_{n-1} \rangle$$
$$= \langle x_n | e^{-\frac{i}{\hbar}(t_n - t_{n-1})H} | x_{n-1} \rangle$$
$$= \langle x_n | e^{-\frac{i}{\hbar} \epsilon H} | x_{n-1} \rangle$$
$$= \frac{1}{2\pi\hbar} \int dp_n \exp\left\{ \frac{i}{\hbar} p_n(x_n - x_{n-1}) - \frac{i}{\hbar} \epsilon H\left(\frac{x_n + x_{n-1}}{2}, p_n \right) \right\}$$

$$(6.15)$$

In which we have identified $x_i \equiv x_0$ and $x_f \equiv x_N$. The above expression is the Feynman's path integral define phase space.

Let's analyze the phase factor more accurately,

$$U(t_f, x_f; t_i, x_i) = \lim_{\substack{\epsilon \to 0 \\ N \to \infty}} \int dx_1 \cdots dx_{N-1} \frac{dp_1}{2\pi\hbar} \cdots \frac{dp_n}{2\pi\hbar} \times$$

$$\times \exp\left\{ \frac{i}{\hbar} \sum_{n=1}^{N} \left[p_n(x_n - x_{n-1}) - \epsilon H\left(\frac{x_n + x_{n-1}}{2}, p_n \right) \right] \right\}$$

$$(6.16)$$

Multiplying and dividing by t, we obtain the expression,

$$\lim_{\substack{\epsilon \to 0 \\ N \to \infty}} \frac{i}{\hbar} \sum_{n=1}^{N} \left[p_n(x_n - x_{n-1}) - \epsilon H\left(\frac{x_n + x_{n-1}}{2}, p_n \right) \right] \quad (6.17)$$

And we observe that in the limit $t \to 0$ the phase factor is proportional to the action of the system.

The most commonly occurring Hamiltonian has the form $\hat{H} = \frac{\hat{P}^2}{2m} + \hat{V}(x)$ a quadratic function of the momentum. The Eq. (6.16) becomes,

$$\lim_{\substack{\epsilon \to 0 \\ N \to \infty}} \frac{i\epsilon}{\hbar} \sum_{n=1}^{N} \left[p_n \frac{x_n - x_{n-1}}{\epsilon} - H\left(\frac{x_n + x_{n-1}}{2}, p_n \right) \right] \quad (6.18)$$

$$= \frac{i}{\hbar} \int_{t_i}^{t_f} dt \left[p\dot{x} - H(x, p) \right]$$

$$= \frac{i}{\hbar} \int_{t_i}^{t_f} dt L \quad (6.19)$$

The integral in the momentum is a Gaussian and can be calculated with easily, obtaining,

$$U(t_f, x_f; t_i, x_i) = \lim_{\substack{\epsilon \to 0 \\ N \to \infty}} \int dx_1 \cdots dx_{N-1} \frac{dp_1}{2\pi\hbar} \cdots \frac{dp_n}{2\pi\hbar}$$

$$\times \exp\left\{ \frac{i\epsilon}{\hbar} \sum_{n=1}^{N} \left[p_n \frac{x_n - x_{n-1}}{\epsilon} - \frac{p_n^2}{2m} - V\left(\frac{x_n + x_{n-1}}{2} \right) \right] \right\}$$

$$(6.20)$$

This relation can be written in a more compact form,

$$\int \frac{dp_n}{2\pi\hbar} e^{-\frac{i\epsilon}{\hbar}\left[\frac{p_n^2}{2m} - p_n \frac{x_n - x_{n-1}}{\epsilon} \right]} = \left(\frac{m}{2\pi i\hbar\epsilon} \right)^{\frac{1}{2}} e^{\frac{im\epsilon}{2\hbar}\left(\frac{x_n - x_{n-1}}{\epsilon} \right)^2} \quad (6.21)$$

obtaining the formal representation

$$U(t_f, x_f; t_i, x_i) = \lim_{\substack{\epsilon \to 0 \\ N \to \infty}} \left(\frac{m}{2\pi i \hbar \epsilon}\right)^{\frac{N}{2}} \int dx_1 \cdots dx_{N-1}$$

$$\times \exp\left\{\frac{i\epsilon}{\hbar} \sum_{n=1}^{N} \left[\frac{m}{2}\left(\frac{x_n - x_{n-1}}{\epsilon}\right)^2 - V\left(\frac{x_n + x_{n-1}}{2}\right)\right]\right\}$$

(6.22)

$$\mathcal{D}x \equiv \lim_{N \to \infty} \prod_{i=1}^{N-1} dx_i \qquad (6.23)$$

We have shown that it is possible to express the time evolution operator in terms of an integral that contains a new integration measure and in named a functional integral.

6.2 INTERPRETATION OF THE CLASSICAL LIMIT

Let us understand the meaning of the new measure $\mathcal{D}x$ and its properties in the functional space of Feynman's paths. Consider the (6.22),the set of points $\{x_0, x_1, ..., x_{N-1}, x_N\}$ can be interpreted as the set of vertices of a set $_N$ such that $_N = \{x(t_i) = x_0, x(t_1) = x_1, ..., x(t_N) = x_N\}$. Then we can say that the integral (6.22) is calculated on all possible elements of $_N$. When $N \to \infty$, $_N$ becomes a continuous curve, $x(t)$, and the integral is calculated on all possible curves with extremes and on all the paths or trajectories that the particle could travel from the starting point $x(t_i)$ to the final $x(t_f)$. Each of them contributes to the transition amplitude with a factor $e^{\frac{i}{\hbar}S[x]}$. Summarizing, our result was obtained from the formulation of quantum mechanics, but this is not the only way to get the (6.23). Indeed Feynman's approach [83] was very different. Based on Dirac's observations he proposed postulates that derive in the equation of Schrödinger and all the apparatus of traditional quantum mechanics:

- the probability that a particle goes from the point x_2 in the instant t_2, to the point x_1 at t_1, is given the module of the transition amplitude $\hat{U}(t_2, x_2; t_1, x_1)$.

- The transition amplitude is obtained from the sum of the contributions of all possible paths connecting the two points at the instants considered (6.23).

- Each path contributes with the same module but with different phase, obtained from the classical action calculated on the path

$$\phi[x(t)] = Ae^{\frac{i}{\hbar}S[x(t)]}. \tag{6.24}$$

And A normalize the U properly.

Contrary to what one can imagine, not all paths contribute with the same weight to the transition amplitude, in fact in all those where $x_n \gg x_{n-1}$ the first term in the exponential of (1.20) becomes very large, and this implies that the corresponding phase oscillates quickly giving an zero average. One of the advantages of Feynman's formulation that the classical limit is very intuitive, as $\hbar \to 0$. In fact, in this situation, every magnitude of system, including the action, is very large compared to $\hbar$, in general if we take two paths, $x_1(t)$ and $x_2(t)$, that are very close, because the variation of the action $|S[x_1] - S[x_2]|$ is small as $|x_1(t) - x_2(t)|$ goes to zero, the phase oscillates rapidly and the contributions of the two paths tend to cancel. The only exception is for all those paths contiguous to the one that exerts the action, that is, $\bar{x}$ satisfying,

$$\left. \frac{\delta S}{\delta x} \right|_{x=\bar{x}} \tag{6.25}$$

and therefore, the classic trajectory is obtained from the principle of minimal action.

It is important to emphasize that in the classical limit is not only the classic trajectory that have the grater contribution, but all the paths close to this also contribute.

6.2.1   Equivalence with Schrödinger's Equation

Let us show how the integral formulation is perfectly consistent with Schrödinger's time-dependent differential equation. In particular, we will retrieve this equation from definition of transition amplitude as a path integral. Consider the propagator between two close instants, obtained in a similar way to (6.25),

$$U(t_f = \epsilon, x_f; t_i = 0, x_i) =$$

$$\left(\frac{m}{2\pi i \hbar \epsilon}\right)^{\frac{1}{2}} \exp\left\{\frac{i\epsilon}{\hbar}\left[\frac{m}{2}\left(\frac{x_f - x_i}{\epsilon}\right)^2 - V\left(\frac{x_f + x_i}{2}\right)\right]\right\} \quad (6.26)$$

and using it to get the wave function at time ϵ with $\psi(x, 0)$,

$$\psi(x, \epsilon) = \int dx' U(\epsilon, x; 0, x') \psi(x', 0)$$

$$= \left(\frac{m}{2\pi i \hbar \epsilon}\right)^{\frac{1}{2}} \int dx' \exp\left[\frac{im}{2\hbar\epsilon}(x - x')^2 - \frac{i\epsilon}{\hbar}V\left(\frac{x + x'}{2}\right)\right]\psi(x', 0)$$

$$(6.27)$$

At this point, by changing the integration variable to $\eta = x' - x$ we get,

$$\psi(x, \epsilon) = \left(\frac{m}{2\pi i \hbar \epsilon}\right)^{\frac{1}{2}} \int d\eta \exp\left[\frac{im}{2\hbar\epsilon}\eta^2 - \frac{i\epsilon}{\hbar}V\left(x + \frac{\eta}{2}\right)\right]\psi(x + \eta, 0)$$

$$(6.28)$$

Since ϵ is very small, the phase η is large and the integrate oscillates quickly around zero without contributing to the result. This suggests that the main contribution is provided by small η and that it is convenient to develop in Taylor's series and neglect the higher order terms η^2 and ϵ, obtaining,

$$\psi(x + \eta, 0) \simeq \psi(x, 0) + \eta\psi'(x, 0) + \frac{\eta^2}{2}\psi''(x, 0),$$

$$e^{-\frac{i\epsilon}{\hbar}V\left(x + \frac{\eta}{2}\right)} = 1 - \frac{i\epsilon}{\hbar}V\left(x + \frac{\eta}{2}\right) + O(\epsilon^2) \simeq 1 - \frac{i\epsilon}{\hbar}V(x) \quad (6.29)$$

Substituting this in (6.28),

$$\psi(x, \epsilon) \simeq \left(\frac{m}{2\pi i \hbar \epsilon}\right)^{\frac{1}{2}} \int d\eta \exp\left(\frac{im}{2\hbar\epsilon}\eta^2\right) \times$$

$$\times \left[\psi(x, 0) + \eta\psi'(x, 0) + \frac{\eta^2}{2}\psi''(x, 0) - \frac{i\epsilon}{\hbar}V(x)\psi(x, 0)\right]$$

calculating the integral,

$$\int d\eta \, \eta e^{i\alpha\eta^2} = 0 \qquad (6.30)$$

$$\int d\eta \, \eta^2 e^{i\alpha\eta^2} = \frac{i}{2\alpha}\sqrt{\frac{i\pi}{\alpha}} \qquad (6.31)$$

it is found,

$$\psi(x,\epsilon) = \psi(x,0) - \frac{i\epsilon}{\hbar}V(x)\psi(x,0) + \frac{i\hbar\epsilon}{2m}\psi''(x,0) \qquad (6.32)$$

Reordering terms we obtain,

$$\frac{\psi(x,\epsilon) - \psi(x,0)}{\epsilon} = -\frac{i}{\hbar}\left[-\frac{\hbar^2}{2m}\frac{\partial^2}{\partial x^2} + V(x)\right]\psi(x,0). \qquad (6.33)$$

In the limit $t \to 0$ the last relation is the time dependent Schrödinger equation.

6.2.2 Composition Rule

It is easily proved that the propagator (6.23) has the following property

$$U(t_2, x_2; t_1, x_1) = \int dx_n U(t_2, x_2; t_n, x_n)U(t_n, x_n; t_1, x_1) , \qquad (6.34)$$
$$\text{with} \quad t_2 > t_n > t_1$$

It is sufficient to consider in the (6.22) the integration variable x_n corresponding to $t = t_n$. The integration on all previous points leads to the propagator $\hat{U}(t_2, x_2; t_n, x_n)$, while the integral on all the subsequent points provide $U(t_n, x_n; t_1, x_1)$. At this point only left the integral on x_n and the result can be written as in (6.34). We also observe that this property is the same as the one presented in (6.11). The interpretation of (6.34) is immediate, if we consider that the amplitude of a transition from point x_1 to x_n and then to x_2 is simply the product of the respective transition amplitudes. These are obtained by adding together the contributions of all possible paths with the considered extremes. Integration over x_n means to account for all trajectories that join extremes x_1 and x_2, as calculating the transition amplitude between the two points.

6.3 FUNCTIONAL INTEGRALS AND PROPERTIES

We begin our presentation discussing the functional derivative generalized to the case of the ordinary functional derivative. This definition is necessary, as we need computational tools that work on functional spaces. A functional and a defined application on a space of functions with values in the real or complex field can be written in the form,

$$F[f] = \int dx\, G(f(x)) \tag{6.35}$$

An example of functional is the action $S[x]$ of the system. Is it possible to define also a functional derivative,

$$\frac{\delta F[f]}{\delta f(y)} = \int dx \frac{\delta G(f(x))}{\delta f(y)} \tag{6.36}$$

with

$$\frac{\delta G(f(x))}{\delta f(y)} = \lim_{\epsilon \to 0} \frac{G(f(x) + \epsilon \delta(x-y)) - G(f(x))}{\epsilon} \tag{6.37}$$

It has the same property of linearity and composition, as an ordinary derivatives. Also we can write the Taylor series of a functional $F[f]$

$$F[f] = F[0] + \int dx_1 \left. \frac{\delta F[f]}{\delta f(x_1)} \right|_{f=0} f(x_1)$$

$$+ \frac{1}{2!} \int dx_1 dx_2 f(x_1) \left. \frac{\delta^2 F[f]}{\delta f(x_1)\delta f(x_2)} \right|_{f=0} f(x_2) + \cdots \tag{6.38}$$

A few examples that will prove useful in the subsequent paragraphs. We calculate the functional derivative of the action using the definition (6.36)

$$S[x] = \int_{t_i}^{t_f} dt'\, L(x,\dot{x}) \quad \text{with} \quad L(x,\dot{x}) = T - V = \frac{1}{2}m\dot{x}^2 - V(x) \tag{6.39}$$

and taking advantage of the simple result,

$$\frac{\delta f(x)}{\delta f(y)} = \delta(x-y) \tag{6.40}$$

First we determine the functional derivative of the kinetic energy T

$$\frac{\delta T(\dot{x}(t'))}{\delta x(t)} = \lim_{\epsilon \to 0} \frac{T\left(\frac{d}{dt'}(x(t') + \epsilon\delta(t' - t))\right) - T(\dot{x}(t'))}{\epsilon}$$

$$= \lim_{\epsilon \to 0} \frac{T\left((\dot{x}(t') + \epsilon\frac{d}{dt'}\delta(t' - t))\right) - T(\dot{x}(t'))}{\epsilon}$$

$$= \frac{\partial T(\dot{x}(t'))}{\partial \dot{x}(t')} \frac{d}{dt'}\delta(t' - t)$$

$$= m\dot{x}(t')\frac{d}{dt'}\delta(t' - t) \tag{6.41}$$

and the potential,

$$\frac{\delta V(x(t'))}{\delta x(t)} = \lim_{\epsilon \to 0} \frac{V(x(t') + \epsilon\delta(t' - t))) - V(x(t'))}{\epsilon}$$

$$= \frac{\partial V(x(t'))}{\partial x(t')}\delta(t' - t)$$

$$= V'(x(t'))\delta(t' - t) \tag{6.42}$$

to get the result,

$$\frac{\delta L}{\delta x(t)} = m\dot{x}(t')\frac{d}{dt'}\delta(t' - t) - V'(x(t'))\delta(t' - t) \tag{6.43}$$

and,

$$\frac{\delta S[x]}{\delta x(t)} = \int_{t_i}^{t_f} dt' \frac{\delta L(x(t'), \dot{x}(t'))}{\delta x(t)} = -m\ddot{x}(t) - V'(x(t)). \tag{6.44}$$

6.3.1 Wick Rotation

A very useful tool to solve problems related to functional integrals is the Wick rotation or Euclidean rotation. Is the analytic continuation of the functional integral to an imaginary time by replacing,

$$t \quad \to \quad t' = -i\tau \quad \text{with } \tau \text{ real} \tag{6.45}$$

A main advantage of this technique involves the regularization of some integrals, because of fluctuations of phase, of the type

$$\int \mathcal{D}x \, e^{\frac{i}{\hbar}S[x]} \tag{6.46}$$

using (6.45) is obtained,

$$
\begin{aligned}
S[x] &= \int_{t_i}^{t_f} dt \left[\frac{m}{2} \left(\frac{dx}{dt} \right)^2 - V(x) \right] \\
&= i \int_{\tau_i}^{\tau_f} d\tau \left[\frac{m}{2} \left(\frac{dx}{d\tau} \right)^2 - V(x) \right] \\
&= iS_E[x]
\end{aligned}
\tag{6.47}
$$

where we have introduced the Euclidean action, defined real and positive,

$$\int_{\tau_i}^{\tau_f} d\tau \left[\frac{m}{2} \left(\frac{dx}{d\tau} \right)^2 - V(x) \right] \tag{6.48}$$

and then the following replacement,

$$I = A \int \mathcal{D}x \, e^{\frac{i}{\hbar}S[x]} \quad \rightarrow \quad I_E = A \int \mathcal{D}x \, e^{-\frac{1}{\hbar}S_E[x]} \tag{6.49}$$

where the second member is well defined.

We note that in the case of Minkowski space-time-dependent functional, Wick rotation allows us to move from the same Minkowski to that Euclidean space. In fact, consider an continuum event $x_\mu = (t, \vec{x})$ with the norm given $x_\mu x^\mu = t^2 - |\vec{x}|^2$. If we carry out a Wick rotation we get $(t, \vec{x}) \rightarrow (i\tau, \vec{x})$ and $t^2|\vec{x}|^2 \rightarrow (t^2 + |\vec{x}|^2)$ that, apart from the negative sign, is the distance in ordinary Euclidean space, hence the name "Euclidean rotation." Finally we look at that following (6.48), the Euler-Lagrange equations describe the motion of a particle in the real Euclidean space subject to a potential $V(x)$.

6.3.2 Semi-Classical Approximation

Let us now discuss one of the most important methods for an approximate calculation of the functional integral. This method is very useful in a vast number of applications. Consider a simplified case, we compute the integral,

$$I = \int_{-\infty}^{+\infty} dx\, e^{-\frac{f(x)}{a}} \tag{6.50}$$

where a and a very small parameter, while f is a function with a single minimum in point x_0. We can expand $f(x)$ around the point x_0, getting,

$$f(x) = f(x_0) + \frac{1}{2!} f''(x_0)(x - x_0)^2 + O\big((x - x_0)^3\big) \tag{6.51}$$

and substituting in the integral we find,

$$I = \int_{-\infty}^{+\infty} dx\, e^{-\frac{f(x_0)}{a} - \frac{f''(x_0)}{2a}(x-x_0)^2 - \frac{O((x-x_0)^3)}{a}} \tag{6.52}$$

Using the substitution $y = \frac{x - x_0}{\sqrt{a}}$, the integral is,

$$I = e^{-\frac{f(x_0)}{a}} \sqrt{a} \int_{-\infty}^{+\infty} dy\, e^{-\frac{f''(x_0)}{2} y^2 - O(\sqrt{a} y^3)} \tag{6.53}$$

and neglecting the last term in the exponential argument leads to the result,

$$I = \sqrt{\frac{2\pi a}{f''(x_0)}} e^{-\frac{f(x_0)}{a}} \tag{6.54}$$

At this point, we note that if the function is at a minimal point must be considered all contributions of these stationary points, so in general the following relationship holds,

$$I = \sum_{min\, x_i} \sqrt{\frac{2\pi a}{f''(x_i)}} e^{-\frac{f(x_i)}{a}} \tag{6.55}$$

This result can also be generalized to a functional integral, and is very helpful in cases where we cannot accurately calculate (6.22). When $\hbar$ is small compared to S is possible to use the above method, and expand the action around the classical trajectory, hence the name semi-classical approximation. The first problem to be addressed is the regularization of the integral, due to its oscillating behavior, through the Wick rotation (6.45). We consider the trajectory $\bar{x}$ that minimizes the action, i.e., the classical trajectory in Euclidean space. Imposing the condition,

$$\frac{\delta S_E}{\delta x(t)}\Bigg|_{x=\bar{x}} = 0 \tag{6.56}$$

Just as before, we expand the action around the trajectory considered,

$$x(t) = \bar{x}(t) + \eta(t) \tag{6.57}$$

where $\eta(t)$ represents the fluctuations around $\bar{x}$, and in this way we get,

$$S_E[x] = S_E[\bar{x}(t) + \eta(t)] = S_E[\bar{x}]$$
$$+ \frac{1}{2} \iint d\tau_1 d\tau_2 \, \eta(\tau_1) \frac{\delta^2 S_E[\bar{x}]}{\delta\bar{x}(\tau_1)\delta\bar{x}(\tau_2)} \eta(\tau_2) + O(\eta^3) \tag{6.58}$$

At this point we insert the previous equation in the integral that appears to the right of (6.49), and perform the replacement (6.57), this leads to the expression, $I_E =$

$$A \int \mathcal{D}x \exp\left(-\frac{S_E[\bar{x}]}{\hbar} - \frac{1}{2\hbar} \iint d\tau_1 d\tau_2 \, \eta(\tau_1) \frac{\delta^2 S_E[\bar{x}]}{\delta\bar{x}(\tau_1)\delta\bar{x}(\tau_2)} \eta(\tau_2) - \frac{O(\eta^3)}{\hbar}\right) \tag{6.59}$$

From this, neglecting the last term in the exponential argument, we get

$$I_E \simeq A e^{-\frac{S_E[\bar{x}]}{\hbar}} \int \mathcal{D}x \exp\left(-\frac{1}{2\hbar} \iint d\tau_1 d\tau_2 \, \eta(\tau_1) \frac{\delta^2 S_E[\bar{x}]}{\delta\bar{x}(\tau_1)\delta\bar{x}(\tau_2)} \eta(\tau_2)\right) \tag{6.60}$$

To calculate the integral above, we can use a suitable basis of orthonormal eigenfunctions in which to expand the fluctuation $\eta(\tau)$. Given that the equality holds,

$$\frac{\delta^2 S_E[\bar{x}]}{\delta\bar{x}(\tau_1)\delta\bar{x}(\tau_2)} = \left(-m\frac{d^2}{d\tau_1^2} + V''(\bar{x})\right)\delta(\tau_1 - \tau_2)$$
$$\equiv \hat{A}(\tau_1)\delta(\tau_1 - \tau_2) \tag{6.61}$$

is convenient to choose the eigenfunctions $\phi_n(\tau_1)$, with eigenvalue λ_n, of the operator $\hat{A}(1)$, and with the boundary conditions $\phi_n(\tau_i) = \phi_n(\tau_f) = 0$.

This way we get,

$$\eta(\tau) = \sum_n c_n \phi_n(\tau) \tag{6.62}$$

We observe that integrate over all possible fluctuations means integrating over all possible values of c_n and therefore we can rewrite (6.60) as follows,

$$I_E \simeq A e^{-\frac{1}{\hbar}S_E[\bar{x}]} \int \prod_n dc_n \exp\left(-\frac{1}{2\hbar}\iint d\tau_1 d\tau_2 \times \right.$$

$$\left. \times \sum_n c_n \phi_n(\tau_1)\hat{A}(\tau_1)\delta(\tau_1 - \tau_2)c_n\phi_n(\tau_2)\right)$$

$$= A e^{-\frac{1}{\hbar}S_E[\bar{x}]} \int \prod_n dc_n \exp\left(-\frac{1}{2\hbar}\sum_n c_n^2 \int d\tau_1\, \phi_n(\tau_1)\hat{A}(\tau_1)\phi_n(\tau_1)\right)$$

$$= A e^{-\frac{1}{\hbar}S_E[\bar{x}]} \int \prod_n dc_n \exp\left(-\frac{1}{2\hbar}\sum_n c_n^2 \lambda_n \int d\tau_1\, \phi_n^2(\tau_1)\right)$$

$$= A e^{-\frac{1}{\hbar}S_E[\bar{x}]} \int \prod_n dc_n \exp\left(-\frac{1}{2\hbar}\sum_n c_n^2 \lambda_n\right)$$

$$= A e^{-\frac{1}{\hbar}S_E[\bar{x}]} \sqrt{\frac{2\pi\hbar}{\prod_n \lambda_n}}$$

$$= \frac{A' e^{-\frac{1}{\hbar}S_E[\bar{x}]}}{\sqrt{\det\left(-m\frac{d^2}{d\tau^2} + V''(\bar{x})\right)}} \tag{6.63}$$

Finally, turning to the ordinary t by performing an inverse Wick rotation,

$$\langle x_f, t_f | x_i, t_i \rangle \simeq \frac{A e^{\frac{i}{\hbar}S[x_{cl}]}}{\sqrt{\det\left(m\frac{d^2}{dt^2} + V''(x_{cl})\right)}} \tag{6.64}$$

The method fails if an eigenvalue of the operator is zero, because, in that case the determinant in (6.64) is zero. In these cases the problem must be approached with more caution. Also note, that if there are more trajectories that minimize the action we must consider all possible contributions, and (6.64) generalizes the as following,

$$\langle x_f, t_f | x_i, t_i \rangle \simeq A \sum_a \frac{e^{\frac{i}{\hbar} S[x_a]}}{\sqrt{\det \left(m \frac{d^2}{dt^2} + V''(x_a) \right)}} \qquad (6.65)$$

6.3.3 Perturbation Theory

In quantum mechanics there are few problems we can solve exactly, and very often, when possible, we use perturbation theory to find an approximate solution. We can also follow this approach with the formalism of functional integration and construct a perturbation theory from which can be obtained a physical interpretation. Consider a particle described initially with a Lagrangian L_0, we know how to calculate exactly the propagator. Suppose that at a time t_i a perturbation described by potential $V(x,t)$ is turned on, where a new Lagrangian is given by $L = L_0 V$. We are interested in the transition amplitude of the particle from the point (t_i, x_i) to (t_f, x_f).

$$U(t_f, x_f; t_i, x_i) = N \int \mathcal{D}x \, \exp\left[\frac{i}{\hbar} \int_{t_i}^{t_f} dt (L_0 - V) \right] \qquad (6.66)$$

If the potential $V(x,t)$ is small compare to $\hbar$, we can construct an exponential expansion,

$$\exp\left[-\frac{i}{\hbar} \int_{t_i}^{t_f} dt \, V(x(t),t) \right]$$
$$= 1 - \frac{i}{\hbar} \int_{t_i}^{t_f} dt \, V(x(t),t) + \frac{1}{2!} \left(-\frac{i}{\hbar} \right)^2 \left[\int_{t_i}^{t_f} dt \, V(x(t),t) \right]^2 + \cdots$$
$$= \sum_{n=0}^{+\infty} \frac{1}{n!} \left(-\frac{i}{\hbar} \right)^n \left[\int_{t_i}^{t_f} dt \, V(x(t),t) \right]^n \qquad (6.67)$$

Introducing the expansion in (6.66), we obtain,

$$U(t_f, x_f; t_i, x_i) = \sum_{n=0}^{+\infty} U_n(t_f, x_f; t_i, x_i) \qquad (6.68)$$

$$U_0(t_f, x_f; t_i, x_i) = \int \mathcal{D}x \, \exp\left[\frac{i}{\hbar} \int_{t_i}^{t_f} dt \, L_0\right] \tag{6.69}$$

$$U_1(t_f, x_f; t_i, x_i) = -\frac{i}{\hbar} \int \mathcal{D}x \, \exp\left[\frac{i}{\hbar} \int_{t_i}^{t_f} dt \, L_0\right] \tag{6.70}$$

$$\int_{t_i}^{t_f} dt_1 \, V(x(t_1), t_1) \tag{6.71}$$

$$U_2(t_f, x_f; t_i, x_i) = \frac{1}{2!} \left(-\frac{i}{\hbar}\right)^2 \int \mathcal{D}x \, \exp\left[\frac{i}{\hbar} \int_{t_i}^{t_f} dt \, L_0\right] \times$$

$$\times \left(\int_{t_i}^{t_f} dt' \, V(x(t'), t')\right)^2 \tag{6.72}$$

For simplicity we omitted the normalization factor N. The propagator the order zero is just what we obtain in the absence of perturbation, while to study the first order correction is appropriate to exchange the two integrations,

$$U_1(t_f, x_f; t_i, x_i) = \int_{t_i}^{t_f} dt_1 \int \mathcal{D}x \left(-\frac{i}{\hbar}\right) V(x(t_1), t_1) \exp\left[\frac{i}{\hbar} \int_{t_i}^{t_f} dt \, L_0\right] \tag{6.73}$$

We observe that for each path the potential depends only of the instant t_1 and position $x(t_1)$ occupied by the particle at that instant. Before and after t_1 the particle wave-function evolves as prescribed by U_0. This prompts us to interpret $\frac{i}{\hbar}V(x(t), t)$ as the amplitude of probability, per unit of time and volume, that the particle interacts with the potential. Then, the rules of composition tell us that the magnitude of the probability from (t_i, x_i) to $(t_1, x(t_1))$, spread by potential, and to arrive to (t_f, x_f) is given by,

$$U_0(t_f, x_f; t_1, x_1) \left(-\frac{i}{\hbar}\right) V(x(t_1), t_1) U_0(t_1, x_1; t_i, x_i) \tag{6.74}$$

Because we need to sum over all possible paths, the (6.74) must be integrate in $x_1 = x(t1)$ and t_1, and the same for (6.34), and obtain (6.71). The second order correction is built in the same way, the particle evolves from (t_i, x_i) to $(t_1, x(t_1))$, goes under an interaction with the potential, evolves from $(t_1, x(t_1))$ to $(t_2, x(t_2))$, with $t_i < t_1 < t2 < tf$, and interacts again, arriving finally in (t_f, x_f). The transition amplitude of this event is,

$$U_0(t_f, x_f; t_2, x_2) \left(-\frac{i}{\hbar}\right) V(x(t_2), t_2) U_0(t_2, x_2; t_1, x_1)$$

$$\times \left(-\frac{i}{\hbar}\right) V(x(t_1), t_1) U_0(t_1, x_1; t_i, x_i) \quad (6.75)$$

Finally if we integrate in $x_1 = x(t_1), t_1, x_2 = x(t_2)$ and t_2 to sum the contributions of all the paths, we obtain,

$$U_2(t_f, x_f; t_i, x_i) = \left(-\frac{i}{\hbar}\right)^2 \int_{t_i}^{t_f} dt_1 \int_{t_1}^{t_f} dt_2 \int dx_1 \int dx_2$$

$$\times V(x(t_2), t_2) V(x(t_1), t_1)$$

$$\times U_0(t_f, x_f; t_2, x_2) U_0(t_2, x_2; t_1, x_1) U_0(t_1, x_1; t_i, x_i)$$

$$= \left(-\frac{i}{\hbar}\right)^2 \int_{t_i}^{t_f} dt_1 \int_{t_1}^{t_f} dt_2$$

$$\int \mathcal{D}x \, V(x(t_2), t_2) V(x(t_1), t_1) e^{\frac{i}{\hbar} S_0[x]} \quad (6.76)$$

Since the integrand $V(x(t_2), t_2) V(x(t_1), t_1)$ is completely symmetric to the interchange of variables t_1 and t_2, results,

$$\int_{t_i}^{t_f} dt_1 \int_{t_1}^{t_f} dt_2 = \frac{1}{2!} \int_{t_i}^{t_f} dt_1 \int_{t_i}^{t_f} dt_2 \quad (6.77)$$

This equation allows us to get exactly (6.72), proving the accuracy of interpretation.

Since the total transition amplitude and is the sum of all contributions, if the series (6.68) converges, the approximation we use to calculate U is better if more diagrams are added together. It is clear that the series expansion obtained from the functional integral that express the propagator U is equivalent to the Dyson series of the time evolution operator.

We observe that in the latter appear explicit ordered products of temporally operators, while Feynman formulation provides automatically the temporal order. We can highlight this property in the simple case of a two-point correlation function between two eigenstates of the operator position. For this, we consider

$$\langle x_f, t_f | X(t_2) X(t_1) | x_i, t_i \rangle \quad \text{with} \quad t_i < t_1 < t_2 < t_f \quad (6.78)$$

and use the completeness of the eigenstates of the operator position into instants t_1 and t_2 in the Heisenberg representation, getting,

$$\langle x_f, t_f | X(t_2)X(t_1)|x_i, t_i \rangle$$

$$= \langle x_f, t_f | X(t_2) \int dx_2|x_2, t_2\rangle\langle x_2, t_2| \int dx_1|x_1, t_1\rangle\langle x_1, t_1| X(t_1)|x_i, t_i\rangle$$

$$= \int dx_2 dx_1 \, x(t_2)x(t_1)\langle x_f, t_f|x_2, t_2\rangle\langle x_2, t_2|x_1, t_1\rangle\langle x_1, t_1|x_i, t_i\rangle \quad (6.79)$$

Since any scalar product in (6.79) can be written as integral on paths, we get,

$$\langle x_f, t_f | X(t_2)X(t_1)|x_i, t_i\rangle = N \int \mathcal{D}x \, x(t_2)x(t_1) \exp\left(\frac{i}{\hbar}\int_{t_i}^{t_f} dt\, L\right)$$
$$(6.80)$$

We observe that $x(t_1)$ and $x(t_2)$ are real quantities and commute, for any $t1$ and t_2, while the first member of the (6.80) is bounded by $t_1 < t_2$. Similarly is proved that we get the same result when the correlation function is consider,

$$\langle x_f, t_f | X(t_1)X(t_2)|x_i, t_i\rangle \quad \text{with} \quad t_i < t_2 < t_1 < t_f \quad (6.81)$$

proving that the following equality is valid,

$$\langle x_f, t_f | T(X(t_2)X(t_1))|x_i, t_i\rangle = N \int \mathcal{D}x \, x(t_1)x(t_2) \exp\left(\frac{i}{\hbar}\int_{t_i}^{t_f} dt\, L\right)$$
$$(6.82)$$

One can extend the result to any set of operators in any state, in each case the path integral presents a measure time ordered.

CHAPTER 7

SUPERSYMMETRY IN QUANTUM MECHANICS

Let us consider particle with mass m in a one-dimensional system. Its temporal evolution is described by Hamiltonian $\hat{H}$ evolving according to Schrödinger equation,

$$i\hbar\frac{\partial|\psi(t)\rangle}{\partial t} = H|\psi(t)\rangle \tag{7.1}$$

Let $V(x)$, the potential energy of the particle as a function of the position. The Hamiltonian operator is,

$$H = \frac{p^2}{2m} + V(x) \tag{7.2}$$

where $\hat{P}$ is momentum operator and $\hat{V}$ the potential energy operator. The Hamiltonian in the representation of the position is,

$$H = -\frac{\hbar^2}{2m}\frac{d^2}{dx^2} + V(x) \tag{7.3}$$

For simplicity and following the literature on the subject we will considered $\hbar = 2m = 1$. The problem that usually arises in the study of a quantum system is the integration of the eigenvalue equation of the Hamiltonian, namely the search for eigenstates and the corresponding eigenvalues of energy. The eigenstates, in the representation of the position, are described by eigenfunctions with the same position variable. Latter they obey the Schrödinger equation independent of time,

$$H\psi(x) = E\psi(x) \tag{7.4}$$

where E represents the eigenvalue of the eigenfunction $\psi(x)$.

Consider systems described by Hamiltonians containing functions of the potential energy that are not singular. In addition, we demand that the energy spectrum contains a discrete part, corresponding to the bound states and that has a minimal value E_{min}. Let $\hat{H}$ be a Hamiltonian with these characteristics, and $V(x)$ the potential energy function. Suppose also, that the minimum value of the energy of the discrete spectrum is 0. This is not a strong condition, since it can be obtained with a shift of all energy levels, rushing the potential $V(x)$ by an additive constant.

The eigenfunction $\psi_0(x)$ associated to the ground state of the system, with eigenvalue of energy corresponding to 0, satisfies the equation,

$$-\psi_0''(x) + V(x)\psi_0(x) = 0 \tag{7.5}$$

where a tilde indicates derivative with respect to x.

Given that the eigenfunction of the ground state, $\psi_0(x)$ must be free of nodes and goes to zero in infinite satisfy the normalization condition. It is possible to establish a relationship between the potential energy function and eigenfunction $\psi_0(x)$, using (7.5),

$$V(x) = \frac{\psi_0''(x)}{\psi_0(x)} \tag{7.6}$$

Knowing the potential, it is possible to derive eigenfunctions related to the ground state, and vice versa, the knowledge of the first eigenfunction allows to determine $V(x)$.

The first step when studying a system through the notion of supersymmetry is to define two operators, $\hat{A}$ and $\hat{A}^\dagger$, which factorize the Hamiltonian,

$$H = A^\dagger A \tag{7.7}$$

The two operators are defined in terms of $W(x)$, called a superpotential, as follows,

$$A = \frac{d}{dx} + W(x), \quad A^\dagger = -\frac{d}{dx} + W(x) \tag{7.8}$$

Defining the momentum operator $\hat{P} = -i\frac{d}{dx}$, we can rewrite the two operators $\hat{A}$ and $\hat{A}$ in the form,

$$A = ip + W(x), \quad A^\dagger = -ip + W(x) \tag{7.9}$$

If $W(x)$ is a real function, $\hat{A}$ and $\hat{A}^\dagger$ are the adjoint of the other. By indicating with E_n the state described by the eigenfunction $\psi_n(x)$, the Schrodinger equation can be written in the form,

$$A^\dagger A \psi_n(x) = E_n \psi_n(x) \tag{7.10}$$

and the condition for $\psi_0(x)$ is,

$$A\psi_0(x) = 0 \tag{7.11}$$

condition enough to ensure that the eigenvalue of the energy relative to the ground state is null.

There is a relationship between the potential $V(x)$ and the superpotential $W(x)$, in fact, from the product,

$$
\begin{aligned}
A^\dagger A &= \left(-\frac{d}{dx} + W(x)\right)\left(\frac{d}{dx} + W(x)\right) \\
&= -\frac{d^2}{dx^2} - W'(x) - W(x)\frac{d}{dx} + W(x)\frac{d}{dx} + W^2(x) \\
&= H
\end{aligned} \tag{7.12}
$$

we get the equation,

$$V(x) = W^2(x) - W'(x) \tag{7.13}$$

which links the superpotential $W(x)$ to the potential $V(x)$. If the wave function of the ground state is known, it is possible to obtain the superpotential directly, from Eq. (7.11), we find,

$$W(x) = -\frac{\psi_0'(x)}{\psi_0(x)} \tag{7.14}$$

The superpotential $W(x)$ is important for defining the supersymmetric partner of the starting potential $V(x)$.

7.1 CONSTRUCTION OF THE SUPERSYMMETRIC PARTNERS

Let $\hat{H}_1$ be the Hamiltonian factorized by the two operators $\hat{A}$ and $\hat{A}^\dagger$ in (7.7). We define the supersymmetrical partner, $\hat{H}_2$, as follows

$$H_2 = AA^\dagger \tag{7.15}$$

or

$$H_2 = -\frac{d^2}{dx^2} + W'(x) + W^2(x) \tag{7.16}$$

The difference from the expression obtained for $\hat{H}_1$ is the sign of the derivative of $W(x)$. A potential $V_2(x)$ can be defined, the supersymmetric partner of $V_1(x)$, as the potential that appears in $\hat{H}_2$:

$$V_2(x) = W^2(x) + W'(x) \tag{7.17}$$

Then, we study the relationships between the energy spectra, and the eigenfunctions of $\hat{H}_1$ and $\hat{H}$. We indicate with $E_n^{(1)}$ and $E_n^{(2)}$ the eigenvalues of the nth excited state, for the Hamiltonians $\hat{H}_1$ and $\hat{H}_2$ respectively, while the eigenfunctions will be denoted with $\psi_n^{(1)}(x)$ and $\psi_n^{(2)}(x)$.

The eigenfunction $\psi_n^{(1)}(x)$ of $\hat{H}_1$ related to the eigenvalue $E_n^{(1)}$ satisfies the equation,

$$H_1 \psi_n^{(1)}(x) = E_n^{(1)} \psi_n^{(1)}(x) \tag{7.18}$$

If we apply the operator $\hat{A}$ to both members of the equation, remembering the fact that $\hat{H}_1$ is defined in (7.7), we get:

$$AA^\dagger A \psi_n^{(1)}(x) = E_n^{(1)} A \psi_n^{(1)}(x) \tag{7.19}$$

That is, through the definition of $\hat{H}_2$,

$$H_2 A\psi_n^{(1)}(x) = E_n^{(1)} A\psi_n^{(1)}(x) \tag{7.20}$$

then $\hat{A}\psi_n^{(1)}(x)$ is an eigenfunction of $\hat{H}_2$ with eigenvalue $E_n^{(1)}$, valid for $n \neq 0$. In the case $n = 0$, the corresponding eigenfunction should be $\psi_0^{(2)}(x) = \hat{A}\psi_0^{(1)}(x)$, but for (7.11) $\psi_0^{(2)}(x)$ is identically zero: there is therefore no state of $\hat{H}_2$ with null eigenvalue.

Similarly, consider the generic eigenfunction $\psi_n^{(2)}(x)$ for a eigenstate of the Hamiltonian $\hat{H}_2$, and with eigenvalue $E_n^{(2)}$, obeying the equation,

$$H_2\psi_n^{(2)}(x) = E_n^{(2)}\psi_n^{(2)}(x) \tag{7.21}$$

applying operator $\hat{A}^\dagger$ to both members, we get,

$$H_1 A^\dagger \psi_n^{(2)}(x) = E_n^{(2)} A^\dagger \psi_n^{(2)}(x) \tag{7.22}$$

This means that by applying the operator $\hat{A}^\dagger$ eigenstates of $\hat{H}_2$, one gets eigenstates of $\hat{H}_1$ with the same eigenvalue. Notice the correspondence between the eigenstates of $\hat{H}_1$ and $\hat{H}_2$, the operator $\hat{A}$ turns eigenstates of $\hat{H}_1$ into eigenstates of $\hat{H}_2$, while the operator $\hat{A}^\dagger$ performs the reverse transformation.

In addition, for both transformations, the energy levels coincide. The only exception to this correspondence is constituted by the ground state of $\hat{H}_1$, which does not correspond to an eigenstate in $\hat{H}_2$. This has important consequences and, as we shall see, it encloses the supersymmetric character of the Hamiltonian $\hat{H}_1$.

As can be seen from Eq. (7.22), it follows that to each eigenstate of $\hat{H}_2$ corresponds a eigenstate of $\hat{H}_1$ obtained by the operator $\hat{A}^\dagger$, whereas for each eigenstate of $\hat{H}_1$ corresponds one of $\hat{H}_2$, except for ground state.

Supposed we have integrated Hamiltonian $\hat{H}_1$, to know its energy spectrum and there exists a bonded state of $\hat{H}_2$, related to the eigenvalue $\bar{E}_n^{(2)}$, and that is not present in spectrum of $\hat{H}_1$. That state is described by eigenfunction $\bar{\psi}_n^{(2)}(x)$. As discussed above, applying to $\bar{\psi}_n^{(2)}(x)$ the operator $\hat{A}^\dagger$ is obtained the eigenfunction of a $\hat{H}_1$ relative to the eigenvalue $\bar{E}_n^{(2)}$. But Since $\hat{H}_1$ is integrated and $\bar{E}_n^{(2)}$ does not appear in its energy spectrum, we obtain an inconsistency, generated by assuming the existence of a state of $\hat{H}_2$ corresponding to a eigenvalue non present between those of

$\hat{H}_1$. We can conclude that $\hat{H}_2$ eigenvalues are only those of $\hat{H}_1$ without the eigenvalue of zero energy: $\hat{H}_2$'s eigenfunctions are only those that can be obtained from those $\hat{H}_1$ through the action of the operator $\hat{A}$.

This result makes possible to affirm that the ground level of the Hamiltonian $\hat{H}_2$ is characterized by the first excited level of $\hat{H}_1$, since no further eigenvalues of energy, in particular negative, are allowed. The first bounded state for $\hat{H}_2$ is that which corresponds to the eigenvalue $E_1^{(1)}$ and the eigenfunction for this state is $\psi_0^{(2)}(x) = \hat{A}\psi_1^{(1)}(x)$. Generalizing this conclusion the nth excited state, we can obtain a relation between the energy spectra of the two partners Hamiltonians,

$$E_0^{(1)} = 0 \tag{7.23}$$

$$E_n^{(2)} = E_{n+1}^{(1)} \tag{7.24}$$

If we consider the eigenfunctions of the two systems, $\hat{A}$ and $\hat{A}^\dagger$ operators transform eigenstate of one Hamiltonian into partner Hamiltonian eigenstates, characterized by the same eigenvalue of energy but with different parity. We can assert that eigenfunctions transforms respectively by creating or destroying a node, and called creation operators and destruction operators to $\hat{A}$ and $\hat{A}^\dagger$ respectively. As with the eigenvalues, we can write the relationship between the two Hamiltonian partners, $\hat{H}_1$ and $\hat{H}_2$,

$$A\psi_{0(1)} = 0 \tag{7.25}$$

$$\psi_n^{(2)} = A\psi_{n+1}^{(1)} \tag{7.26}$$

$$\psi_{n+1}^{(1)} = A^\dagger\psi_n^{(2)} \tag{7.27}$$

These relations work modulo a normalization constant. A notion of wave functions for the states of a system must be recalled: the quantum number n, indicates the excitation level, this corresponds to the number of nodes of the function, so a leap from a certain level to an excited state corresponds to an extra node. Furthermore, while increasing in number, the nodes are positioned in intermediate positions between the pre-existing ones. For a symmetrical potential with respect to the source, the wave functions will be parity defined with respect to the inversion of the coordinate. The number of nodes of a function allows to determine its parity: an odd number of nodes characterizes as an antisymmetric function, while a even number of nodes is a symmetric one. The state of a system can be characterized as bosonic or fermionic depending on whether the wave function is even or odd, respectively.

7.2 ALGEBRA OF A SUPERSYMMETRIC SYSTEM

Considering Hamiltonian $\hat{H}_1$ and his supersymmetric partner $\hat{H}_2$ obtained by (7.15), we can define a Hamiltonian $\hat{H}$,

$$H = \begin{pmatrix} H_1 & 0 \\ 0 & H_2 \end{pmatrix} = \begin{pmatrix} A^\dagger A & 0 \\ 0 & AA^\dagger \end{pmatrix} \tag{7.28}$$

in a Hilbert space where the generic state vector $|\psi_n$ is defined by following,

$$|\psi_n\rangle = \begin{pmatrix} \psi_n^{(1)} \\ \psi_n^{(2)} \end{pmatrix} \tag{7.29}$$

The energy level corresponding to the ground state is eigenvalue only of $\hat{H}_1$, so it does not degenerate for the Hamiltonian $\hat{H}$; all other eigenvalues are degenerate twice, since they correspond to a eigenstates of $\hat{H}_1$ and $\hat{H}_2$. We define two other operators, indicated by $\hat{Q}$ and $\hat{Q}^\dagger$, which are called supercharges,

$$Q = \begin{pmatrix} 0 & 0 \\ A & 0 \end{pmatrix}, \quad Q^\dagger = \begin{pmatrix} 0 & A^\dagger \\ 0 & 0 \end{pmatrix} \tag{7.30}$$

that determines the rules that define the algebra of a supersymmetric system. The two operators $\hat{Q}$ and $\hat{Q}^\dagger$ commute with the Hamiltonian $\hat{H}$:

$$[Q, H] = [Q^\dagger, H] = 0 \tag{7.31}$$

Since $\hat{Q}$ and $\hat{Q}^\dagger$ do not explicitly depend on time, this property expresses the supercharges conservation in time, as we can see from Heisenberg's equation of motion for a generic operator $\hat{O}$,

$$\frac{d\mathcal{O}}{dt} = \frac{\partial \mathcal{O}}{\partial t} - i[\mathcal{O}, H] \tag{7.32}$$

The $\hat{Q}$ and $\hat{Q}^\dagger$ operators are also nilpotent,

$$\{Q, Q\} = \{Q^\dagger, Q^\dagger\} = 0 \tag{7.33}$$

In fact, considering the anti-commutator between the two operators, is valid the equation

$$\{Q, Q^\dagger\} = \{Q^\dagger, Q\} = H \tag{7.34}$$

A pair of these operators, along with Hamiltonian $\hat{H}$, satisfying the rules (7.31), (7.33) and (7.34) defines a supersymmetric system. It is interesting to give an interpretation of this formalism. Consider the Hilbert space $\mathcal{H}$ on which Hamiltonian $\hat{H}$ acts, as the direct product of two spaces. The first is that generated by the Hamiltonian eigenstates, in this case $\hat{H}_1$, The second is a space whose base is formed by the eigenstates of an operator called the fermionic number. A space in which each state corresponds to a well-defined number of characteristic entities is called a Fock's space. In this case, a Fock space generated by the eigenstates of the $\hat{N}_f$ operator, and is defined as follows,

$$N_f = \frac{1}{2}(1 - \sigma_3) = \frac{1}{2}\left[\begin{pmatrix} 1 & 0 \\ 0 & 1 \end{pmatrix} - \begin{pmatrix} 1 & 0 \\ 0 & -1 \end{pmatrix}\right] = \begin{pmatrix} 0 & 0 \\ 0 & 1 \end{pmatrix} \tag{7.35}$$

where σ_3 indicates the third Pauli matrix or σ_z. The eigenvalues of $\hat{N}_f$ are 0 and 1: the eigenstates of the operator are those where there are fermions in the system, or there are none. In the ket representation they are expressed as,

$$|0\rangle = \begin{pmatrix} 1 \\ 0 \end{pmatrix} \qquad |1\rangle = \begin{pmatrix} 0 \\ 1 \end{pmatrix} \tag{7.36}$$

We can define operators that increase or decrease the number of fermions in the system, acting on the eigenstates. They are respectively,

$$\psi = \begin{pmatrix} 0 & 1 \\ 0 & 0 \end{pmatrix} \qquad \psi^\dagger = \begin{pmatrix} 0 & 0 \\ 1 & 0 \end{pmatrix} \tag{7.37}$$

Observe that,

$$\{\psi, \psi\} = \{\psi^\dagger, \psi^\dagger\} = 0 \tag{7.38}$$

The two operators are nilpotent, when applied more than once to an eigenstate, it generates a null ket, regardless of the starting state. This is a formalization of Pauli's exclusion principle: the fermionic number cannot be increased if it is equal to one, or be decremented if it is zero.

The eigenstates of the Hamiltonian $\hat{H}$ are obtained as the direct product between the eigenstates of $\hat{H}_1$ and those of the fermionic space, whereby the generic ket $|\psi\rangle$ can be written in the form,

$$|\psi\rangle = |n\rangle \otimes |n_f\rangle \tag{7.39}$$

where n represents the quantum number for excitations of $\hat{H}_1$. Each energy level is doubly degenerate because there are two states, one bosonic, with $n_f = 0$, and one fermionic, with $n_f = 1$, which corresponds to the same energy. The passage from one to the other is called supersymmetry transformation, and is achieved by the operators $\hat{Q}$ and $\hat{Q}^\dagger$. Degeneration does not occur in the case of the ground level. In fact, according to the hypothesis already made on $\hat{H}_1$, the Hamiltonian $\hat{H}$ is definite positive, so the expectation value for any positive ket is,

$$\langle \psi | H | \psi \rangle \geq 0 \tag{7.40}$$

Recalling the relation (7.34), we can rewrite (7.40) as follows,

$$\langle \psi | QQ^\dagger + Q^\dagger Q | \psi \rangle \geq 0 \tag{7.41}$$

or

$$|Q^\dagger |\psi\rangle|^2 + |Q|\psi\rangle|^2 \geq 0 \tag{7.42}$$

In the case of the ground state $|\psi\rangle$, the expectation value coincides with the eigenvalue and is zero, then, the first member of inequality (7.42) is zero only if both terms are zero, so they must be,

$$Q|0\rangle = Q^\dagger|0\rangle = 0 \tag{7.43}$$

The ground state therefore breaks the degeneration that characterizes all the other levels. Considering the eigenstates of $\hat{H}$, it can give a different characterization from that expressed in the definition (7.39): to associate a state with each of the energy levels, so that they are fermionic or bosonic. Of those in the fermionic state there may be at most one, while the bosonic state can be occupied by a large number: the Fock's space states are distinguished by two quantum numbers, the number of bosons and fermions present in the system.

The generic state vector can be expressed in the form $|n_b, n_f\rangle$, where n_b and n_f are the bosonic and fermionic numbers respectively. The number n_b can be any integer value, while n_f can take only 0 or 1 values.

Through the creation and destruction operators $\hat{\psi}$ and $\hat{\psi}^\dagger$ in the fermionic it is possible to factorize the operators $\hat{Q}$ and $\hat{Q}^\dagger$,

$$Q = A\psi^\dagger, \qquad Q^\dagger = A^\dagger \psi \tag{7.44}$$

To explain the action of operators $\hat{Q}$ and $\hat{Q}^\dagger$ on a ket of the space $\mathcal{H}$, consider these as ket of Fock's space identified by the fermionic and bosonic numbers. The operators ψ and $\psi^\dagger$ act on the fermionic number of a generic eigenstate as they act on the space created by the operator $\hat{N}_f$ while $\hat{A}$ and $\hat{A}^\dagger$ act on the states of the Hamiltonian $\hat{H}_1$, the latter increase the quantum number excitation thus varying The bosonic number, decreasing or increasing in a unit, respectively. The action of $\hat{Q}$ on a generic state $|n_b, n_f\rangle$ can therefore be expressed as,

$$\begin{aligned}
Q|n_b, n_f\rangle &= A\psi^\dagger |n_b, n_f\rangle \\
&= (A \otimes \mathbb{I}_f) \left(\mathbb{I} \otimes \psi^\dagger \right) |n_b, n_f\rangle
\end{aligned} \tag{7.45}$$

in which $\mathbb{I}_f$ and $\mathbb{I}$ indicate the unit matrices respectively in the fermion space and in the space of the energy eigenstates. This notation is justified as the space we are considering is given as the tensorial product between the space of the fermionic number and the space of energy eigenstates, it follows that an operator acting on the tensor product state, has to composed of two operators, one acting on the first space and one on the second.

The action of operators $\hat{A}$ and $\hat{A}^\dagger$ results in decreasing or increasing, respectively, the quantum number n_b, so that we can write,

$$Q|n_b, n_f\rangle = A\psi^\dagger |n_b, n_f\rangle = |n_b - 1, n_{f+1}\rangle \tag{7.46}$$

Thus, it is possible to describe the action of the operator $\hat{Q}$ remembering that $\hat{A}$ decreases the quantum level of a $\hat{H}_1$, and hence the quantum number associated with it. The energy of the system remains unchanged since the fermionic number is increased.

The state of the system is described in the space $\hat{N}_f$ by the vector $|1\rangle$. Applying $\hat{Q}^\dagger$ obtains the ket $|0\rangle$, that is, the fermionic state is eliminated, but the action of the other operator $\hat{A}$ causes the creation of a bosonic state: a fermionic state is transformed into a bosonic. This is possible as any energy eigenvalue relative to an excited state is doubly degenerate, and each corresponds to a fermionic and a bosonic eigenstate.

Applying again the $\hat{Q}^\dagger$ operator to the resulting state we obtain the null ket, also, applying the fermionic decreasing operator on the state $|0\rangle$ gives the null vector, so the whole tensor product is zero.

Now suppose the particle in the ground state, and it is a bosonic, since the wave function does not have zeros and is symmetrical with respect to the origin. The ground state is then described in the space $\hat{N}_f$ of the vector $|0\rangle$. Applying the operator $\hat{Q}$, the vector $|1\rangle$ is obtained, that is, the boson state is transformed into a fermion one. However, this operation produces the zero vector due to (7.43), so the eigenvalue of the ground energy is not degenerate, and is bosonic.

7.3 VIOLATION OF THE SUPERSYMMETRY

If we consider the Hamiltonian $\hat{H}$ defined in the expression (7.28), and let $|0\rangle$ be the ket of the space $\mathcal{H}$ relative to the ground state, the $\hat{Q}$ and $\hat{Q}^\dagger$ supercharges are also defined. The supersymmetry is said to be non-violating if the following relations are hold,

$$Q|0\rangle = Q^\dagger|0\rangle = 0 \tag{7.47}$$

The relation (7.47) expresses the fact that energy relative to the ground state is a non-degenerate value relative to a bosonic state of the system. In fact the ground state is,

$$|0\rangle = \begin{pmatrix} \psi_0^{(1)} \\ 0 \end{pmatrix} \tag{7.48}$$

so the Hamiltonian $\hat{H}$ verify this relation.

We can verify this condition or its violation by asking, given a known superpotential $W(x)$, if it is possible to construct a supersymmetrical Hamiltonian. The condition to be noted of the superpotential function is,

$$|W(x)| \to \infty, \qquad \text{when} \quad x \to \pm\infty \qquad (7.49)$$

This, in order to obtain a discrete spectrum and the possibility of finding normalized eigenfunctions. Starting from $W(x)$, the $\hat{A}$ and $\hat{A}^\dagger$ operators can be constructed, and the two Hamiltonians $\hat{H}_1$ and $\hat{H}_2$ can be constructed also, using their respective potential definitions (7.13) and (7.17). To check if at least one of the two Hamiltonians has a null eigenstate, and then consider the other as a supersymmetric partner, let us take the equations separately

$$\frac{d\psi_0^{(1)}}{dx}(x) + \psi_0^{(1)}(x)W(x) = 0$$
$$\frac{d\psi_0^{(2)}}{dx}(x) - \psi_0^{(2)}(x)W(x) = 0 \qquad (7.50)$$

from the two equations (7.50), we obtain the solutions,

$$\psi_0^{(1,2)}(x) = N \exp\left(\mp \int^x W(t)dt\right) \qquad (7.51)$$

where the minus sign provides a solution to the first equation, and the plus sign of the second. Since solutions of the equation are wave functions for the ground state with zero eigenvalue, they need to be normalizable. Because the system is supersymmetrical it is necessary that at least one of the two solutions is a normalizable function. Suppose that the function $\psi_0^{(1)}(x)$ is normalizable, so the system formed by the Hamiltonian $\hat{H}_1$ and $\hat{H}_2$ is supersymmetrical, the energy spectrum of $\hat{H}_1$ contains the eigenvalue for the null state, while for $\hat{H}_2$ the eigenvalue of the ground level coincides with the first excited level of $\hat{H}_1$. In this case supersymmetry is said to be not violated.

Supersymmetry is said to be violated if the Hamiltonian does not allow a null eigenstate of energy, i.e., none of the two solutions of (7.51) is a normalizable function. In this case, however, it is possible to determine the $\hat{A}$ and $\hat{A}^\dagger$ operators, but they do not act as in the case of non-violated supersymmetry. To have an example of superpotential for which the supersymmetrical character of the built system depends on the values assumed by some parameters, consider

$$W(x) = gx^n \tag{7.52}$$

The parameters g and n determine the supersymmetric character of the Hamiltonian system

$$H_1 = -\frac{d^2}{dx^2} + gx^{2n} + ngx^{n-1}$$

$$H_2 = -\frac{d^2}{dx^2} + gx^{2n} - ngx^{n-1} \tag{7.53}$$

To find out if there is a normalizable solution to the Schrödinger equation for the ground state with eigenvalue zero, the expression (7.51) is used. Analyzing the function $\psi_0^{(1)})(x)$ is obtained,

$$\frac{d}{dx}\psi_0^{(1)}(x)\frac{1}{\psi_0^{(1)}(x)} = -gx^n \tag{7.54}$$

$$\ln[\psi_0^{(1)}(x)] = -\frac{g}{n+1}x^{n+1} + c \tag{7.55}$$

$$\psi_0^{(1)}(x) = c' \exp\left[-\frac{g}{n+1}x^{n+1}\right] \tag{7.56}$$

If n and g are positive, the function $\psi_0^{(1)})(x)$ is zero in the limit $x \to +\infty$ but diverges if $x \to -\infty$. Changing the sign of g reverse the results of the previous limits, but we still do not get a normalizable function. Considering, however, n odd and g positive the function is normalizable, because it is even, and tends to zero with $x \to \pm\infty$. A similar results would be obtained by considering $\psi_0^{(2)})(x)$, imposing n odd and g negative. Thus, it is possible to determine the ket for the ground state of the $\hat{H}$ system, replacing the expression obtained from (7.54) in (7.48).

If we consider a Hamiltonian, $\hat{H}_1$, factorized by the operators $\hat{A}$ and $\hat{A}^\dagger$s, and construct the partner Hamiltonian $\hat{H}_2$ as pointed out in (7.7) and (7.15). If the eigenvalue of energy at the ground level is not null, the occurs the supersymmetry breaking. The condition imposed by Eq. (7.11) fails because,

$$A^\dagger A\psi_0^{(1)}(x) = E_0^{(1)}\psi_0^{(1)}(x) \tag{7.57}$$

If we apply the operator $\hat{A}$ to both members of the equation above, by using the definition of $\hat{H}_2$ (7.15) it results,

$$H_2 A \psi_0^{(1)}(x) = E_0^{(1)} A \psi_0^{(1)}(x) \tag{7.58}$$

Note that Hamiltonian $\hat{H}_2$ admits the eigenvalue of energy $E_0^{(1)}$, while in the case of non-broken supersymmetry it was missing. One can consider the generic eigenstate of $\hat{H}$ described by the eigenfunction $\psi_n^{(1)}(x)$ with eigenvalue $E_n^{(1)}$, and note that this eigenstate corresponds to one of $\hat{H}_2$. Similarly, considering the generic eigenfunction $\psi_n^{(2)}(x)$ for a eigenstate of $\hat{H}_2$, if $E_n^{(2)}$ is the corresponding eigenvalue, we get the equation,

$$H_1 A^\dagger \psi_0^{(1)}(x) = E_0^{(1)} A^\dagger \psi_0^{(1)}(x) \tag{7.59}$$

It is concluded that, as in the case of non-broken supersymmetry, operators A and A establish a bi-univocal relationship between the eigenfunctions of $H1$ and $H2$, and relate eigenfunctions to the same eigenvalues of energy. The energy spectra of the two Hamiltonians coincide, and since both have the same eigenvalue for the ground state, the relation between the corresponding eigenvalues is,

$$E_0^{(1)} \neq 0 \tag{7.60}$$
$$E_n^{(2)} \neq E_n^{(1)} \tag{7.61}$$

Corresponding eigenvalues are characterized by the same quantum number, that is, they refer to equally excited eigenstates. This is the difference between a system that violates supersymmetry and one that respects it. It rises from the observation that both Hamiltonians are characterized by the same eigenvalue for ground level. This conclusion drops an important element of the supersymmetrical systems, i.e., the possibility to link systems, both bosonic and fermionics, characterized by the same level of energy.

The way that the relation between the eigenvalues is changed, also modifies that between eigenfunctions: modulo a normalization constant, is shown by the equations (7.58) and (7.59) that,

$$A\psi_0^{(1)} \neq 0 \qquad (7.62)$$

$$\psi_n^{(2)} = A\psi_n^{(2)} \qquad (7.63)$$

$$\psi_n^{(1)} = A^\dagger \psi_n^{(2)} \qquad (7.64)$$

It will be shown how to proceed with the study of a Hamiltonian in the event that there is a discreet and bounded spectrum of energy, but unknown. Let $\tilde{V}(x)$ be a generic potential, not singular, and E_n and $\psi_n(x)$, the energy spectrum and eigenfunction of the corresponding Hamiltonian. Let $\phi(x)$ be a solution, not necessarily eigenfunction, of the Schrödinger equation, with eigenvalue ϵ. The equation under consideration is,

$$-\phi''(x) + \tilde{V}(x)\phi(x) = \epsilon\phi(x) \qquad (7.65)$$

The only condition that is imposed on (x) is to be free of nodes, and is annulled for $x \to -\infty$. We define the superpotential $W_\phi(x)$ as follows,

$$W_\phi(x) = -\frac{\phi'(x)}{\phi(x)} \qquad (7.66)$$

Then follows the definition of the potential associated with $W_\phi(x)$,

$$
\begin{aligned}
V_{1\phi}(x) &= W_\phi(x)^2 - W_\phi'(x) \\
&= \frac{\phi'^2(x)}{\phi^2(x)} - \frac{\phi'^2(x) - \phi''(x)\phi(x)}{\phi^2(x)} \\
&= \frac{\phi''(x)}{\phi(x)} \\
&= \tilde{V}(x) - \epsilon \qquad (7.67)
\end{aligned}
$$

the sub-index 1 serves to distinguish from the potential $V_{2\phi}(x)$,

$$V_{2\phi}(x) = W_\phi(x)^2 + W_\phi'(x) \qquad (7.68)$$

The eigenvalues of the Hamiltonian containing $V_{1\phi}(x)$ are given in the expression obtained in (7.67),

$$E_{n\phi} = E_n - \epsilon \qquad (7.69)$$

Three cases will arise, depending on whether the difference $E_0 - \epsilon$, or $E_{n\phi}$, is positive, zero, or negative. In that only the first two cases will be considered, since the third implies nodes in the function (x), i.e., singularities in the superpotential and potentials.

If the difference is strictly positive, the system violates supersymmetry, in fact $W_\phi(x)$ can be seen as a superpotential that generates a system that does not allow a zero ground state of energy or degenerate, and therefore supersymmetry is broken.

If the difference is zero, the superpotential considered generates a supersymmetric system, since $E_0 = \epsilon$, it must necessary that the function from which it originates coincides with eigenfunction $\psi_0(x)$.

The supersymmetric character of a system can be determined by the value of a parameter, known as Witten's index. In order to calculate it, we need to know the spectrum eigenstate of the system. Considering the fermionic number $\hat{N}_f$, hereafter referred to as F, the Witten's index is defined as,

$$\Delta = \text{Tr}[(-1)^F] \tag{7.70}$$

Since $(-1)^F = e^{i\pi}$ we have,

$$
\begin{aligned}
(-1)^F &= \exp[i\pi F] \\
&= \sum_{k=0}^{+\infty} \frac{(i\pi)^k F^k}{k!} = \mathbb{I} + \sum_{k=1}^{+\infty} \frac{(i\pi)^k F^k}{k!} \\
&= \mathbb{I} + \sum_{k=1}^{+\infty} \frac{(i\pi)^k F}{k!} = \mathbb{I} + F \sum_{k=0}^{+\infty} \frac{(i\pi)^k}{k!} - F \\
&= \mathbb{I} - 2F = \begin{pmatrix} 1 & 0 \\ 0 & -1 \end{pmatrix}
\end{aligned}
\tag{7.71}
$$

We obtain an operator with eigenvalues $+1$ and -1. Applying this operator to a system state, we determine parity. Considering two states, a fermionic $|f\rangle$ and a bosonic $|b\rangle$, the following equalities are valid,

$$(-1)^F |f\rangle = -|f\rangle \tag{7.72}$$

$$(-1)^F |b\rangle = |b\rangle \tag{7.73}$$

To calculate the trace of the operator, the average values are calculated for each possible state,

$$\Delta = \sum_k \langle k|(-1)^f|k\rangle \qquad (7.74)$$

In the previous expression $|k\rangle$ is an energy eigenstate with two components, related to the corresponding fermionic and bosonic states.

Considering a generic state where both components of the ket $|k\rangle$ are not null, the expectation value of $(1)^F$ is null, for a system formed by two Hamiltonian partners, the expectation value for any state other than ground is necessarily zero, because the energy spectra coincide, and therefore the correspondence between eigenstates must be valid. The contribution of the Witten index comes from the operator's expectation value for the ground state: in the case of broken supersymmetry it is zero, since for both Hamiltonians there is a state with the same minimum energy, then $\Delta = 0$. In the Supersymmetry case, however, the expectation value is one since the second component of the ket $|0\rangle$ is zero. We can then define the values that Witten's index can have: zero in case of violated supersymmetry, $+1$ in the case of non broken supersymmetry.

The meaning that can be attributed to the integer Witten's index, provides the difference between the number of states of the two partner systems: in the case of a violation of such supersymmetry, is zero, while it is different from zero in cases where supersymmetry is verified. By indicating with $N_+(E = 0)$ the number of bosonic states with zero energy and $N_-(E = 0)$ the corresponding number of fermion states, the Witten index is,

$$\Delta = N_+(E = 0) - N_-(E = 0) \qquad (7.75)$$

7.4 SUCCESSIVE FACTORING OF A HAMILTONIAN

In (7.1) we considered a Hamiltonian $\hat{H}_1$ having a discrete and bounded spectrum, indicated by $E_0^{(1)}$, the energy of the ground state defines the superpotential function $W(x)$ and through this, the operators $\hat{A}$ and $\hat{A}^\dagger$ so that results in,

$$H_1 = A^\dagger A + E_1^{(0)}$$
$$= -\frac{d^2}{dx^2} + W^2(x) - W'(x) + E_1^{(0)}$$
$$= -\frac{d^2}{dx^2} + V_1(x) \tag{7.76}$$

Next, the partner supersymmetric Hamiltonian $\hat{H}_2$ was built,

$$H_2 = AA^\dagger + E_1^{(0)}$$
$$= -\frac{d^2}{dx^2} + W^2(x) + W'(x) + E_1^{(0)}$$
$$= -\frac{d^2}{dx^2} + V_2(x) \tag{7.77}$$

Recalling the relation (7.14), writing $\overset{(1)}{_0}(x)$ as the wave function of the ground state of $\hat{H}_1$, we can express $V_2(x)$ in the form

$$V_2(x) = V_1(x) + 2W'(x)$$
$$= V_1(x) - 2\frac{d^2}{dx^2}\ln\left(\psi_0^{(1)}(x)\right) \tag{7.78}$$

We want to repeat the same procedure, applying it to the Hamiltonian $\hat{H}_2$, with $\hat{A}_i$ and $\hat{A}_i^\dagger$ will now denote the factoring operators of the i-th Hamiltonian, and with $W_i(x)$ its related superpotential. For example, operators $\hat{A}$, $\hat{A}^\dagger$ and $W(x)$ will be indicated with $\hat{A}_1$, $\hat{A}_1^\dagger$ and $W_1(x)$ respectively. Considering the relationship (7.23), the energy of the ground level of $\hat{H}_2$ is,

$$E_0^{(2)} = E_1^{(1)} \tag{7.79}$$

Suppose therefore that we can construct the superpotential function $W_2(x)$ and deduct from it operators $\hat{A}_2$ and $\hat{A}_2^\dagger$ such that,

$$H_2 = A_2^\dagger A_2 + E_2^{(0)}$$
$$= -\frac{d^2}{dx^2} + W_2^2(x) + W_2'(x) + E_1^{(1)}$$
$$= -\frac{d^2}{dx^2} + V_2(x) \tag{7.80}$$

Indicating with $\psi_0^{(2)}(x)$ the wave function for the ground estate of Hamiltonian $\hat{H}_2$, it can be define,

$$
\begin{aligned}
V_3(x) &= W_2^2(x) + W_2'(x) + E_1^{(1)} \\
&= V_2(x) + 2W_2'(x) \\
&= V_2(x) - 2\frac{d^2}{dx^2}\ln\left(\psi_0^{(2)}(x)\right) \\
&= V_1(x) - 2\frac{d^2}{dx^2}\ln\left(\psi_0^{(1)}(x)\right) - 2\frac{d^2}{dx^2}\ln\left(\psi_0^{(2)}(x)\right) \\
&= V_1(x) - 2\frac{d^2}{dx^2}\ln\left(\psi_0^{(1)}(x)\psi_0^{(2)}(x)\right)
\end{aligned}
\tag{7.81}
$$

It follows the definition of the supersymmetrical partner of $\hat{H}2$,

$$
\begin{aligned}
H_3 &= A_2 A_2^\dagger + E_1^{(1)} \\
&= -\frac{d^2}{dx^2} + V_3(x)
\end{aligned}
\tag{7.82}
$$

The relationship between the eigenvalues of energy is,

$$
E_n^{(3)} = E_{n+1}^{(2)} = E_{n+2}^{(1)}
\tag{7.83}
$$

Thus, the factorization of Hamiltonian $\hat{H}_2$ has generated another characterized by the same discrete spectrum but lacking of the corresponding ground state, $\hat{H}_3$ have an eigenvalue of energy of the ground state, that corresponds to the second excited level of $\hat{H}_1$. Through the superpotential $W_2(x)$, operators $\hat{A}_2$ and $\hat{A}_2^\dagger$ can be defined by applying the definition (7.8), with these, it is possible to obtain the eigenfunction of the Hamiltonian $\hat{H}_3$ from those of $\hat{H}_1$, in fact, besides a normalization constant it is observed that,

$$
\begin{aligned}
\psi_n^{(3)}(x) &= A_2 \psi_{n+1}^{(2)}(x) \\
&= A_2 A_1 \psi_{n+2}^{(1)}(x)
\end{aligned}
\tag{7.84}
$$

The subsequent factorization process for a Hamiltonian can be repeated for each discrete energy level, thus leading to the construction of a partner family of Hamiltonians, where from the energy spectrum the first $m-1$ eigenvalues of $\hat{H}_1$ is absent.

We can generalize the relation (7.81), (7.83) and (7.84), written for the potentials, with eigenvalues and the eigenfunction of the first three Hamiltonians, obtaining,

$$V_m(x) = V_1(x) - 2\frac{d^2}{dx^2}\ln\left(\psi_0^{(m-1)}(x)\cdots\psi_0^{(1)}(x)\right) \tag{7.85}$$

$$\psi_n^{(m)} = A_{m-1}\cdots A_1\psi_{n+m-1}^{(1)}(x) \tag{7.86}$$

$$\psi_n^{(1)} = A_1^\dagger\cdots A_{n-1}^\dagger\psi_0^{(n)}(x) \tag{7.87}$$

$$E_n^{(m)} = E_{n+m-1}^{(1)} \tag{7.88}$$

Suppose we want to study eigenfunctions for the bound states of a Hamiltonian with a below-bounded discrete spectrum: the generic nth-excited level can be studied by building the Hamiltonian $\hat{H}_n$ and calculating its eigenfunction of the ground state. This operation may sometimes be simpler than the direct integration of Hamiltonian, since the calculation of a ground eigenfunction ensures that the solution found is free of nodes.

CHAPTER 8

QUANTUM PERSISTENT CURRENTS

8.1 PERSISTENT CURRENTS

Persistent currents are those, which, as its name states, do not change in time. It is a quantum effect revealed in microscopic and mesoscopic systems, in which a charged particle moves around a close geometry, such as a ring, in the presence of a magnetic flux. Specifically. This current is produced for any ring configuration, since the ground state for the electron in the ring is always nonzero. This is due to the shift produced by a phase introduced as a consequence of the magnetic flux, an Aharanov-Bohn phase. The calculation of the persistent current in a ring with several electrons can be reduced from the calculation of the current associated with the energy of an electron, summing over these currents. The simplest model is a metal ring, and a current generated by a single electron.

The system consists in an one-dimensional microscopic ring with a singe electron with mass m and charge $e-$, and where the boundary conditions of wave functions are periodic. Parameterizing the ring of longitude L with the coordinate u. The time independent Schrödinger equation is given by,

$$\frac{-\hbar^2}{2m}\frac{\partial^2 \psi_n(u)}{\partial u^2} = \epsilon_n \psi_n(u), \tag{8.1}$$

where the function $\psi_n(u)$ meets the periodicity condition:

$$\psi_n(u + L) = \psi_n(u).$$

The eigenfunctions $\psi_n(u)$ are,

$$\psi_n(u) = \frac{1}{\sqrt{L}} exp\left(2\pi in \frac{u}{L}\right), \qquad (8.2)$$

and its eigenenergies ϵ_n are,

$$\epsilon_n = \frac{h^2}{2mL^2} n^2. \qquad (8.3)$$

where n is the principal quantum number.

Now, we consider the effects of applying a constant uniform magnitude field $\vec{B}$ to the system. We choose the coordinates as shown in the Figure 8.1, where the magnetic field is parallel to the z axis, being perpendicular to the plane of the ring. The magnitude field is constrained in such a way that it does not directly interact with the ring, but the potential Vector associated with this does it,

When a charged particle of mass m and charge e is in the presence of a magnitude field, the momentum changes from to being **P** to $\mathbf{P}mec - e\vec{A}$, where $\mathbf{P}mec = mv$ is the mechanic moment of the particle associated with the speed **v** and $\vec{A}$ es the potential vector associated with the magnetic field that satisfies the condition $\mathbf{B}\nabla \times \vec{A}$. Then the Hamiltonian becomes,

$$\mathbf{H} = \frac{\left(\mathbf{P}_{mec} - e\vec{A}\right)^2}{2m}.$$

For $\vec{B} = B\hat{z}$ we can define $\vec{A} = \frac{Br}{2}\hat{\theta}$. Also, we can write $\vec{A} = \frac{\phi}{L}\hat{\theta}$, where $\phi = \frac{L^2}{4\pi}B$ is the flow enclosed by the ring. The coordinate $u = \frac{L\theta}{2\pi}$ follows the circumference of the ring, so that the derivative d/du is always tangent to the trajectory and therefore parallel to $\hat{\theta}$.

With this in mind, the Schrödinger equation independent of time is written as:

$$\frac{1}{2m}\left(-i\hbar\frac{d}{du} - e\frac{\phi}{L}\right)^2 \psi_n(u) = \epsilon_n\psi_n(u). \qquad (8.4)$$

The eigenfunctions are given again by the Eq. (8.4), while the eigenenergies are given by,

$$\epsilon_n = \frac{h^2}{2mL^2} \left(n - \frac{\phi}{\phi_0} \right)^2, \tag{8.5}$$

here we introduce the quantum flux $\phi_0 = \dfrac{h}{e}$.

As can be seen, for $n = 0$ the energy is nonzero, for it always has a nonzero angular momentum, which gives rise to the perpetual motion of the particle in the ring.

Another approach to the issue of persistent currents can be though using a gauge transformation. In quantum mechanics a gauge transformation is a change in the potential vector $A \to A + \nabla\xi$, this will involve a change in the wave function associated,

$$\psi' = \psi e^{-i\frac{\xi}{\hbar}} \tag{8.6}$$

where ξ depends in general on the position. Since $\nabla \times \nabla\xi = 0$ regardless the function ξ, this change is not reflected on the magnetic field, given by $B = \nabla \times A$.

The effect on the Hamiltonian will be on the kinetic term,

$$(\hat{P} + e\hat{A})\psi' = (-i\hbar\nabla)\psi' + \left(a\hat{A} + e\nabla\xi \right) \psi' \tag{8.7}$$

$$= \left(-i\hbar e^{-i\frac{e\xi}{\hbar}} \nabla\psi - e(\nabla\xi)\psi' \right) + \left(e\hat{A} + e\nabla\xi \right) \psi' \tag{8.8}$$

$$= e^{-i\frac{e\xi}{\hbar}} \left(-i\hbar\nabla + e\hat{A} \right) \psi \tag{8.9}$$

$$= e^{-i\frac{e\xi}{\hbar}} \left(\hat{P} + e\hat{A} \right) \psi \tag{8.10}$$

$$\tag{8.11}$$

The transformation carries over a phase in the Hamiltonian, leaving unchanged the wave function. Both approaches have similar physical consequences. Let us take as a specific example the gauge given by $\nabla\hat{A} = 0$, called Lorentz condition. This fixed the ξ by the relation

$\nabla \hat{A}' = 0 \rightarrow \nabla(\hat{A} + \xi) = 0$ giving $-\nabla\hat{A} = \nabla\xi$. Then,

$$\xi(r) = -\int_{r_0}^{r} dr' \hat{A}(r') - n \oint dr' \hat{A}(r'), \qquad (8.12)$$

where the first integral is independent of the path of integration and the second, count the n times the ring encircle.

Using the Stoke's theorem,

$$n \oint dr' \hat{A}(r') = n\phi \qquad (8.13)$$

where ϕ is the magnetic flux enclosed by the path or the total flux threading the ring. Under the gauge transformation, the periodicity condition of the wave functions is,

$$\psi(u + L) = e^{\frac{e}{\hbar}i\xi(u+L)} e^{-\frac{e}{\hbar}i\xi(u)} \psi(u) \qquad (8.14)$$

but, because a full circling in the ring corresponds to $n = 1$, the term $\xi(u + L)$ is,

$$\xi(r + L) = -\int_{r_0}^{r} dr' \hat{A}(r') - \oint dr' \hat{A}(r') \qquad (8.15)$$

Substituting accordantly in (8.14),

$$\psi(u + L) = e^{\frac{e}{\hbar}i\left(-\int_{r_0}^{r} dr' \hat{A}(r') - \oint dr' \hat{A}(r')\right)} e^{\frac{e}{\hbar}i\left(\int_{r_0}^{r} dr' \hat{A}(r')\right)} \psi(u)$$

$$= e^{\frac{e}{\hbar}i\left(-\oint dr' \hat{A}(r')\right)} \psi(u) \qquad (8.16)$$

which is,

$$\psi(u + L) = e^{\frac{e}{\hbar}in\phi} \psi(u) = e^{i2\pi n \frac{\phi}{\phi_0}} \psi(u). \qquad (8.17)$$

8.2 PERSISTENT CHARGE CURRENTS

To calculate a charge persistent current, we find the current associated with each eigenstate. An electron, with charge $-e$ that moves at a velocity v along the ring, takes a time $\Delta t = L/v$ to traverse the total length of the ring (L being the ring circumference length), then, the average current is the charge passing through any point on the ring per unit time in a period, $i = -e/\Delta t = -ev/L$. The velocity of a charged particle in a magnetic field can be written as $v = P_{mechanic}/m = (P_{can} + eA)/m$. In the system we have been considering, the velocity v_n associated with the n-state satisfies,

$$
\begin{aligned}
v_n \psi_n(u) &= \frac{(P_{can} + eA)}{m} \psi_n(u) \\
&= \frac{1}{m} \left(-i\hbar \frac{d}{du} + e\frac{\phi}{L} \right) \left(\frac{1}{\sqrt{L}} exp \left(2\pi in \frac{u}{L} \right) \right) \\
&= \frac{h}{mL} \left(n + \frac{\phi}{\phi_0} \right) \frac{1}{\sqrt{L}} exp \left(2\pi in \frac{u}{L} \right),
\end{aligned}
\tag{8.18}
$$

then we have the current,

$$
\begin{aligned}
i_n &= \frac{-ev_n}{L} \\
&= -\frac{eh}{mL^2} \left(n + \frac{\phi}{\phi_0} \right).
\end{aligned}
\tag{8.19}
$$

It follows that is possible to write the current in terms of the Eq. (8.5) as,

$$
i_n = -\frac{\partial \epsilon_n}{\partial \phi}.
\tag{8.20}
$$

Now consider the case of a ring with several electrons, we assume for simplicity that electrons do not interact with each others. To find the total current I at a temperature T, we add all the contributions of each level to a weighting factor $f(\epsilon, \mu, T)$, given by Fermi-Dirac's distribution function,

$$
\begin{aligned}
I &= \sum_n i_n f(\epsilon_n, \mu, T) \\
&= \left(1 + exp \left(-\frac{(\mu - \epsilon_n)}{K_B T} \right) \right)^{-1} \\
&= -\frac{\partial \Omega}{\partial \phi},
\end{aligned}
\tag{8.21}
$$

where $\Omega = -K_{BT} \sum_n \left(ln \left(1 + exp \left(\frac{(\mu - \epsilon_n)}{K_{BT}} \right) \right) \right)$ is the grand canonical thermodynamical potential, where the index n also encompasses both states of spin. We assume that the temperature is well below the Fermi temperature, so that the potential $\mu \approx \epsilon_F$, where ϵ_F is the Fermi energy.

In the low-temperature limit $T \to 0$, the Fermi-Dirac distribution $f(\epsilon_n, \epsilon_F, T)$ becomes step function, or Heaviside $\Theta(\epsilon_F - \epsilon_n)$. Thus, the sum of the Eq. (??) becomes a sum over the currents of all energy levels below the Fermi level.

The expression for the persistent charge current is written,

$$I_c(\phi) = \int_{-\infty}^{\mu} J_c(E) dE, \tag{8.22}$$

where $J_c(E)$ is the charge density current and is given by,

$$J_c = \frac{e}{N} \mathbf{V}. \tag{8.23}$$

8.2.1 Temperature Effects in the Persistent Charge Current

For many electrons, the relation is also valid, provided that there is no effective interaction between them. This assumption is especially appropriate in the case of metals since the long-range interactions are suppressed by the charge screening. At some temperature T each of the contributions to the current is weighted by the Fermi-Dirac distribution $f(\epsilon_n, \epsilon_F, T)$,

$$I = \sum_n f(\epsilon_n, \epsilon_F, T) i_n, \tag{8.24}$$

where ϵ_F is the Fermi level.

$$I = -\sum_n \left(1 + e^{-\frac{(\epsilon_F - \epsilon_n)}{k_B T}} \right)^{-1} \frac{\partial \epsilon_n}{\partial \Phi}, \tag{8.25}$$

and n includes now the distinction for each spin state. When $T \to 0$ then $f(\epsilon_n, \epsilon_F, T) \to \Theta(\epsilon_F - \epsilon_n)$, the Heaviside function truncates the levels in a sum of single-level currents.

The currents are the slopes of the for each energy level in Figure 8.1. Each of the slopes of successive levels are anti-correlated, the current of each electron level added to the ring will tend to cancel the contribution of the previous level, resulting in a current essentially dominated by the highest energy level.

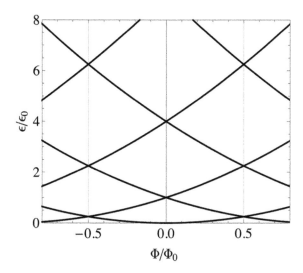

Figure 8.1: Energy as a function of the magnetic flux in dimensionless units; $\epsilon_0 = \frac{h}{2mL^2}$.

.

8.3 PERSISTENT SPIN CURRENTS

It is well known the growing impact of spintronics in condensed matter due to its possible applications to technology. Spintronic devices are based on large changes in the resistance, as a spin-polarized current is pass through an interface where the polarization changes abruptly. The creation of more complex devices that include the spin degree of freedom as the fundamental, dynamical entities has been proposed, as the Datta and Das transistor [26]. Naturally a spin current is the expected quantity to be controlled. Actually this currents are constructed by means of ferromagnets that polarize the charge current, although is a simple solution, the construction of such devices at reduced scales has shown to be rather difficult. Alternative ways are being studied, such as the introduction of materials that interact with the spin directly and so the transport of spin can be manipulated.

A pure spin current is characterized for the absence of net charge transport. This can be quantified intuitively in a simple expression,

$$J_s = I_\uparrow - I_\downarrow, \tag{8.26}$$

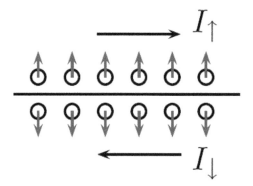

Figure 8.2: Intuitive definition of a pure spin current.

the arrows representing the spin polarization of the charge currents.

A charge current flows in one direction, carrying a spin polarization. Another current flows in the reverse direction with carrying opposite spin polarization. The charge contribution to the total current vanishes. The spin polarization on the other hand will endure, manifesting as a magnetic accumulation in both edges of the sample (Figure 8.2).

$$J_c = I_\uparrow - I_\downarrow = 0 \qquad (8.27)$$

$$J_s = I_\uparrow + I_\downarrow = 2I_\uparrow. \qquad (8.28)$$

It is then ambiguous given this definition whether it is possible to measure a spin current directly or not, but through the spin accumulation.

8.3.1   Spin Currents and Spin Hall Effect

The term "spin Hall effect" was introduced by Hirsch in 1999 [122]. It is indeed similar to the normal Hall effect, where charges of opposite signs accumulate at the sample boundaries due to the Lorentz force in a magnetic field. But there are significant differences. First, there is no need of magnetic field for spin accumulation, if a magnetic field perpendicular to the spin direction is applied, it will destroy the spin polarization. Second, the value of the spin polarization at the boundaries is limited by spin relaxation, and the polarization exists in relatively wide spin layers determined by the spin diffusion length, typically on the order of $1\mu m$. It originates from the coupling between charge and spin currents due to the SO interaction. It consists in a spin accumulation at the lateral boundaries of a current conductor, the direction of the spin are opposite at both boundaries.

The spin current can be introduced by a tensor J_i^a, where the a index is for the spin orientation and the index i is for the flow direction. If it is a pure current in the x direction, polarized in z, the only component different from zero is J_x^z. Under space inversion both spin and charge currents change sign, in contrast under time reversal inversion the charge current changes sign while spin current is symmetric. A straightforward phenomena where two charge currents fulfill the conditions for spin current generation is the spin hall effect.

The standard way to define spin currents is through the equation,

$$J_i^a = \frac{1}{2} n \left\langle \sigma^a v_i + v_i \sigma^a \right\rangle, \tag{8.29}$$

which is basically the difference of the charge carriers currents with opposite spin, with n the charge density. The spin Hall conductivity is defined

$$\sigma_{SH} = -\frac{J_i^a}{E_j}, \tag{8.30}$$

and it measures the conductivity due to a small electric field E_j applied.

It is tempting to describe spin transport in a scheme similar to the charge transport theory. Because of charge conservation, charge densities ρ_c and charge currents J_c satisfy the continuity equation $\rho_c + \nabla J_c = 0$. For spin transport, we can consider the spin density S_i instead of ρ_c. The Definition (8.29) is the natural generalization of the charge model, nonetheless, spin-orbit coupling violates spin conservation, and the continuity equation for spin densities and currents does not hold. The spin orbit interaction can be seen in a simple picture, consisting in an interpretation as a Zeeman term, which in the case of Rashba is $-\sigma \cdot B_{Rashba}^{eff.} = -\sigma \cdot (\alpha p_\phi e_\rho)$ and the precession axis of the spin rotates as the electron moves along. The Eq. (8.29) as definition of the spin current can be used as long as its limitations permit. Despite the fact that it cannot be directly related to spin accumulation, it is a useful model quantity to compare the effect of different spin-orbit coupling mechanisms. The continuity equation does not hold, but for instance, for some models it is possible to evaluate source terms [27], which are often termed as spin torque. Other definitions of spin currents have also been proposed. Analyzing the current of the total angular momentum $L_z + S_z$ [28] arguing that it vanishes for the Rashba Hamiltonian H_R due to the rotational invariance of H_R and that thus the presence or not of with impurities would determine angular momentum currents. A definition that is not proportional to $\langle \sigma^a v_i + v_i \sigma^a \rangle$ was proposed by Shi et al.[29] , is given as the time derivative of a "spin displacement."

8.4 EQUILIBRIUM CURRENTS IN A MESOSCOPIC RING COUPLED TO A RESERVOIR

It is of special interest that SO may be used to implement control of the spin degree of freedom, since it weakly couple to decoherence effects compared to the electric charge, moreover the topological effects, such as Aharonov-Bohm, Aharanov-Casher [24, 51], and Hall effects works as a protection mechanism for charge and spin currents, reducing the electronic dispersion and even emulating superconductivity.

8.5 THE DECOUPLED SO ACTIVE RING

Consider a 2DEG into a ring in polar coordinates. The expression for the Rashba Spin Orbit (RSO) potential in Cartesian coordinates is,

$$\tilde{V}_R = \alpha(\sigma_x \Pi_y - \sigma_y \Pi_x), \tag{8.31}$$

where α is the coupling strength, σ_i denotes the Pauli matrices and $\Pi_i = p_i - eA_i$ where A_i are the components of the vector potential associated with an external magnetic field in the $\hat{z}$ direction. The potential is evidently Hermitian.

A straight-forward coordinate change $(x, y) \rightarrow (\rho, \phi)$ results in a non-Hermitian form that must be symmetrized appropriately. The correct Hermitian RSO potential in polar coordinates is given by the usual coordinate transformation plus a basis rotation of the spinor [52],

$$V_R = e^{i\sigma_z \frac{\varphi}{2}} \tilde{V}_R e^{-i\sigma_z \frac{\varphi}{2}} = -\hbar \omega_{SO} \sigma_\rho \left(i\partial_\varphi + \frac{\Phi}{\Phi_0} \right) - i\hbar \frac{\omega_{SO}}{2} \sigma_\varphi, \tag{8.32}$$

where $\omega_{SO} = \frac{\alpha \hbar}{a}$, a is the ring radius and $\Phi_0 = 2\pi\hbar/e$ is the quantum of flux. The rotated Pauli matrices are defined as $\sigma_\varphi = -\sigma_x \sin\varphi + \sigma_y \cos\varphi$

and $\sigma_\rho = \sigma_x \cos\varphi + \sigma_y \sin\varphi$. Adding the kinetic energy operator reads the Hamiltonian

$$H = \hbar\Omega \left(i\frac{\partial}{\partial\varphi} + \frac{\Phi}{\Phi_0} \right)^2 - \hbar\omega_{SO}\sigma_\rho \left(i\frac{\partial}{\partial\varphi} + \frac{\Phi}{\Phi_0} \right) - i\frac{\hbar\omega_{SO}}{2}\sigma_\varphi, \quad (8.33)$$

with $\Omega = \hbar/2ma^2$. Carefully completing squares taking into account operator ordering and the angular dependencies of σ_φ and σ_ρ, one arrives at the compact form,

$$H = \hbar\Omega \left(-i\frac{\partial}{\partial\varphi} - \frac{\Phi}{\Phi_0} + \frac{\omega_{SO}}{2\Omega}\sigma_\rho \right)^2 - \frac{\hbar\omega_{SO}^2}{4\Omega}. \quad (8.34)$$

In order to obtain the eigenvalues we focus only on the quadratic term, and restore the additive scalar term to the resulting eigenvalue. We can then solve the simpler eigenvalue equation

$$\left(-i\frac{\partial}{\partial\varphi} - \frac{\Phi}{\Phi_0} + \frac{\omega_{SO}}{2\Omega}\sigma_\rho \right)\psi = \sqrt{\frac{E}{\hbar\Omega}}\psi, \quad (8.35)$$

clearly ψ, is an eigenfunction of the square of the previous operator with the square of the eigenvalue. The proposed form for the eigenspinor is

$$\psi_j^\mu(\varphi) = e^{in_j^\mu\varphi}\chi^\mu(\varphi) = e^{in_j^\mu\varphi} \begin{pmatrix} A^\mu \\ e^{i\varphi}B^\mu \end{pmatrix} \quad (8.36)$$

where j labels right and left propagating plane waves ($j = 1$ clockwise and $j = 2$ counterclockwise), μ is the spin label and $n_j^\mu \in \mathbb{Z}$ ($\mu = 1$ spin up and $\mu = 2$ spin down). Solving the matrix equation, the eigenvalues are found to be,

$$E_{n,j}^\mu = \hbar\Omega \left((-1)^j n + \frac{1}{2\pi}\Phi_{AB} - \frac{1}{2\pi}\Phi_{AC}^{(\mu)} \right)^2 - \frac{\hbar\omega_{SO}^2}{4\Omega} \quad (8.37)$$

were $\Phi_{AB} = \Phi/\Phi_0$, using AB for Aharonov-Bohm phase, and

$$\Phi_{AC} = \pi(1 + (-1)^\mu \sqrt{1 + (\omega_{SO}/\Omega)^2}) \quad (8.38)$$

AC for Aharonov-Casher phase. The eigenfunction coefficients satisfy the relation

$$\frac{\Omega}{\omega_{SO}} \left(1 + (-1)^\mu \frac{1}{\cos\theta} \right) A^\mu = B^\mu, \tag{8.39}$$

with $\cos\theta = 1/\sqrt{1 + (\omega_{SO}/\Omega)^2}$. One can then choose $A^{(1)} = B^{(2)} = \cos\frac{\theta}{2}$ and $-A^{(2)} = B^{(1)} = \sin\frac{\theta}{2}$. We thus arrive at the eigenfunctions The magnetic flux breaks the time reversal symmetry (TRS) associated with the simultaneous change of j and μ.

$$\psi_j^1(\varphi) = e^{in_j^1\varphi} \begin{pmatrix} \cos\frac{\theta}{2} \\ e^{i\varphi}\sin\frac{\theta}{2} \end{pmatrix},$$

$$\psi_j^2(\varphi) = e^{in_j^2\varphi} \begin{pmatrix} \sin\frac{\theta}{2} \\ -e^{i\varphi}\cos\frac{\theta}{2} \end{pmatrix}, \tag{8.40}$$

where $\frac{\theta}{2} = \tan^{-1}(\Omega/\omega_{SO} - \sqrt{(\Omega/\omega_{SO})^2 + 1})$.

The spin-orbit interaction alone preserves time reversal symmetry, so in the absence of a magnetic field $E_{n,+}^\uparrow = E_{n,-}^\downarrow$. At half integer flux quanta this degeneracy is repeated. For other values of the flux the degeneracy is broken. For zero SO coupling and in the absence of a Zeeman term there is a peculiar two fold degeneracy for each level due to the closing of the wave function for half integer spin. Thus $E_{n,-}^\uparrow = E_{n+1,-}^\downarrow$ and $E_{n,+}^\downarrow = E_{n+1,+}^\uparrow$ for all fluxes. At zero and half integer flux quanta we have four-fold degeneracy in the absence of SO coupling. Such degeneracies are important when computing the corresponding charge and spin currents.

8.6   DECOHERENCE WITH SPIN ORBIT COUPLING

Buttiker introduced an ingenious way to couple a simple quantum system to reservoir that behaved like a voltage probe. As the coupling to the reservoir is not defined in Hamiltonian terms and leads to dephasing, we have a Hamiltonian solution to the uncoupled problem and a scattering approach for the coupling to the reservoir. The two problems meet when using the

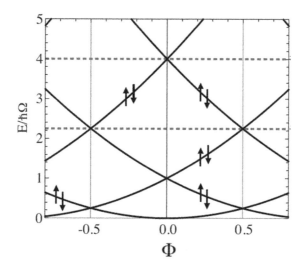

Figure 8.3: Energy states of the decoupled ring. The two-fold degeneration is present. $(\omega_{SO}/\Omega = 0)$. The energy is expressed in units $E_0 = \hbar\Omega$. The two dashed lines represent Fermi levels considered to compute the charge and spin currents.

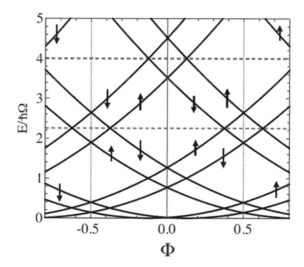

Figure 8.4: Energy states of the decoupled ring. $(\omega_{SO}/\Omega = 0.75)$. The energy is expressed in units $E_0 = \hbar\Omega$. The two dashed lines represent Fermi levels considered to compute the charge and spin currents.

complete basis of the uncoupled problem with the coefficients of these basis components determined by matching boundary conditions.

Coupling to the reservoir is introduced in a ring through an ideal lead that acts as a voltage probe. The reservoir emits electrons with a Fermi distribution and absorbs electrons of any energy. Dephasing occurs due to the absence of a phase relation between injected and emitted electrons at a particular energy.

The coupling between the lead and the ring is described by the dispersion matrix S which relates the incoming and outgoing amplitudes $\vec{a}' = S\vec{a}$. The current conservation implies that S is unitary, the matrix is 3×3 one for each spin label μ, as the coupling to the reservoir is spin symmetric. Each entry corresponds to transmission (t) and reflections (r) from the lead to the ring, and from the ring to the lead. In general the matrix S will dependent on five independent parameters, considering S to be symmetric with respect to the two branches of the ring, the number of independent parameters reduces to three.

$$S = \begin{pmatrix} r_{33} t_{32} t_{31} \\ t_{23} r_{22} t_{21} \\ t_{13} t_{12} r_{11} \end{pmatrix} = \begin{pmatrix} -(a+b)\sqrt{\varepsilon}\sqrt{\varepsilon} \\ \sqrt{\varepsilon} \quad a \quad b \\ \sqrt{\varepsilon} \quad b \quad a \end{pmatrix}, \qquad (8.41)$$

where $a = (\sqrt{1-2\varepsilon}-1)/2$, $b = (\sqrt{1-2\varepsilon}+1)/2$ and ε is the coupling parameter with the reservoir, which varies between 0 and $1/2$ for the uncoupled and fully coupled limits respectively[53]. The symmetry of the terms in the S matrix depends on which fields are present, as we will see below.

The lead coupling the ring to the reservoir needs two equivalent spin channels and thus can be expanded as

$$\psi_{lead}(x) = \sum_{\mu=1,2} \phi_{lead}(x)\chi^{(\mu)}(0)x \in (-\infty, 0]. \qquad (8.42)$$

where x is the coordinate along the lead and $x = 0$ is defined as the coordinate at which the lead connects to the ring, while the reservoir is at $x = -\infty$, and χ^μ is a two component spinor eigenstate of the σ_z. As the lead is not spin-orbit active the energies are $E = \hbar^2 k^2/2m$. The coefficients of the expansion in (8.42) are given by,

$$\phi_{lead}(x) = \sqrt{\mathcal{N}} \left(e^{ikx} + C_3 e^{-ikx}\right). \tag{8.43}$$

The normalization pre-factor is determined considering an energy interval $E, E + dE$, the differential of current injected into the lead is $dI = ev(dN/dE)f(E)dE$, where $f(E)$ is the Fermi distribution, $dN/dE = 1/2\pi\hbar v$ is the density of states of a perfect lead, and $v = \hbar k/m$. The wave function for the lead contemplates the correct current if $\mathcal{N} = f(E)dE/2\pi\hbar v$.

For the ring wave function, it is now a mixture of the four basis functions of the uncoupled case, so that we may accommodate for the new boundary conditions, we define

$$\Psi(\varphi) = C_1^1 \psi_1^1(\varphi) + C_1^2 \psi_1^2(\varphi) + C_2^1 \psi_2^1(\varphi) + C_2^2 \psi_2^2(\varphi) \tag{8.44}$$

The coefficients are to be fixed by imposing equality of the wave functions at $x = 0$ for $\varphi = 0$ and 2π. The dispersion problem is written as $\vec{a}'^{(\mu)} = S\vec{a}^{(\mu)}$, the coefficients $\vec{a}^{(\mu)} = (\alpha^{(\mu)}, \beta^{(\mu)}, \gamma^{(\mu)})$ and $\vec{a}'^{(\mu)} = (\alpha'^{(\mu)}, \beta'^{(\mu)}, \gamma'^{(\mu)})$ are found evaluating (8.43) at the junction at $x = 0$ for $\alpha^{(\mu)}$ and $\alpha'^{(\mu)}$, and evaluating ψ_2^μ in $\varphi = 0, 2\pi$ for the $\beta'^{(\mu)}$ and $\gamma^{(\mu)}$, respectively. The coefficients $\beta^{(\mu)}$ and $\gamma'^{(\mu)}$ evaluating ψ_1^μ in $\varphi = 0, 2\pi$ respectively. The set of equations can be cast, for each spin subspace as

$$\begin{pmatrix} \sqrt{\mathcal{N}} C_3^\mu \\ C_1^\mu \\ C_2^\mu \end{pmatrix} = \begin{pmatrix} -(a+b) & \sqrt{\varepsilon} & \sqrt{\varepsilon} e^{2\pi i n_1^\mu} \\ \sqrt{\varepsilon} & a & b e^{2\pi i n_1^\mu} \\ \sqrt{\varepsilon} e^{-2\pi i n_2^\mu} & b e^{-2\pi i n_2^\mu} & a e^{-2\pi i (n_1^\mu - n_2^\mu)} \end{pmatrix} \begin{pmatrix} \sqrt{\mathcal{N}} \\ C_2^\mu \\ C_1^\mu \end{pmatrix}. \tag{8.45}$$

where we have absorbed the phase factors into a redefined S matrix that manifestly displays the symmetry of the system: Note that we can invert for the quantum number as a function of the energy and fields

$$n_j^\mu = (-1)^j \sqrt{\frac{E}{\hbar\Omega}} + \frac{\Phi}{\Phi_0} - \frac{1}{2}\left(1 + (-1)^\mu \sqrt{1 + \left(\frac{\omega_{SO}}{\Omega}\right)^2}\right). \tag{8.46}$$

Referring to (8.41) one can readily check that, in the absence of magnetic or SO fields, $t_{jk} = t_{kj} = \sqrt{\varepsilon} e^{2\pi i n_1^\mu} = \sqrt{\varepsilon} e^{-2\pi i n_2^\mu}$, i.e., S is an orthogonal (symmetric) matrix, time reversal invariant. When the magnetic field is on but there is no SO coupling, then $t_{jk} \neq t_{kj}$ so $n_1^\mu \neq n_2^\mu$ and time reversal symmetry is broken. When the magnetic field is turned off and the SO coupling is present, time reversal symmetry is restored, and there is the additional symmetry for changing j and μ labels simultaneously. Thus the larger 6×6 matrix $S \otimes \mathcal{I}_s$ matrix is symplectic and embodies Kramer's degeneracy.

Solving the system of equations one can obtain each of the amplitudes

$$C_1^\mu = \frac{\sqrt{\epsilon \mathcal{N}} \left(1 - e^{2\pi i n_2^\mu}\right)}{\left(1 - be^{2\pi i n_1^\mu}\right)\left(b - be^{2\pi i n_2^\mu}\right) + a^2 \left(1 - be^{2\pi i n_1^\mu}\right)},$$

$$C_2^\mu = \frac{\sqrt{\epsilon \mathcal{N}} \left(e^{2\pi i n_1^\mu} - 1\right)}{\left(1 - be^{2\pi i n_1^\mu}\right)\left(b - be^{2\pi i n_2^\mu}\right) + a^2 \left(1 - be^{2\pi i n_1^\mu}\right)},$$

$$C_3^\mu = \frac{\epsilon \left(e^{2\pi i n_1^\mu} - 1 + \left(1 - e^{2\pi i n_2^\mu}\right) e^{2\pi i n_1^\mu}\right)}{\left(1 - be^{2\pi i n_1^\mu}\right)\left(b - be^{2\pi i n_2^\mu}\right) + a^2 \left(1 - be^{2\pi i n_1^\mu}\right)}$$
$$- (a + b). \tag{8.47}$$

For the charge density the modulus squared of the coefficients acquire a particularly simple form in terms of the coupling parameters,

$$|C_1^\mu|^2 = \frac{2\epsilon \mathcal{N}}{g^{(\mu)}} \left(1 - \cos\left(2\pi n_2^\mu\right)\right), \tag{8.48}$$

$$|C_2^\mu|^2 = \frac{2\epsilon \mathcal{N}}{g^{(\mu)}} \left(1 - \cos\left(2\pi n_1^\mu\right)\right), \tag{8.49}$$

$$|C_3^\mu|^2 = 1, \tag{8.50}$$

where

$$g^{(\mu)} = 3 + \sqrt{1 - 2\varepsilon} - 3\varepsilon - 2\left(1 + \sqrt{1 - 2\varepsilon} - \varepsilon\right)\cos\left(2\pi n_1^\mu\right) +$$
$$+ 2\sqrt{1 - 2\varepsilon}\cos\left(2\pi\left(n_1^\mu - n_2^\mu\right)\right) - 2\cos\left(2\pi n_2^\mu\right) + \cos\left(2\pi\left(n_1^\mu + n_2^\mu\right)\right) +$$
$$+ \left(\sqrt{1 - 2\varepsilon} - \varepsilon\right)\left(-2\cos\left(2\pi n_2^\mu\right) + \cos\left(2\pi\left(n_1^\mu + n_2^\mu\right)\right)\right). \tag{8.51}$$

For the density of states (DOS) we know that the number of electron in the energy interval dE is given by $dN = \left|C_1^1\right|^2 + \left|C_1^2\right|^2 + \left|C_2^1\right|^2 + \left|C_2^2\right|^2$.

As each amplitude modulus is proportional to the energy interval dE and using the chain rule $dN/dk = (dN/dE)(dE/dk) = (dN/dE)\hbar^2 k/m$. The number of electrons per unit energy range is given by

$$\frac{dN}{dE} = \sum_{i,\mu} \frac{\varepsilon f(E)}{\pi \hbar v} \frac{\left(1 - \cos 2\pi n_i^{\mu}\right)}{g^{(\mu)}}, \tag{8.52}$$

Then the DOS is given by,

$$\frac{dN}{dk} = \frac{2\varepsilon}{\pi} \left(\frac{\sin^2 \left(2\pi n_1^1\right) + \sin^2 \left(2\pi n_2^1\right)}{g^{(1)}} + \frac{\sin^2 \left(2\pi n_1^2\right) + \sin^2 \left(2\pi n_2^2\right)}{g^{(2)}} \right). \tag{8.53}$$

The explicit relation between DOS and energy comes from substituting the expressions for $n_j^{\mu} = (-1)^j \sqrt{E/\hbar\Omega} + \Phi/\Phi_0 - 1/2 \left(1 + (-1)^\mu \sqrt{1 + (\omega_{SO}/\Omega)^2}\right)$ from the uncoupled problem. These expressions now define this quantum number that becomes a continuous function of the energy and flux and SO coupling, no longer restricted to be integer or half integer, as the problem is coupled.

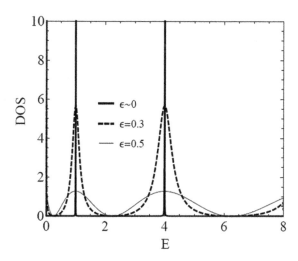

Figure 8.5: The density of states on the ring as a function of the energy for two values of the reservoir coupling parameter. ($\omega_{SO}/\Omega = 0$, $\Phi/\Phi_0 = 0$). The energy is expressed in units $E_0 = \hbar\Omega$ and the Fermi level is $E_F = 8E_0$.

The limit of zero fields (neither SO nor magnetic field) with coupling to the reservoir recovers Buttiker's result[84],

$$
\frac{dN}{dk} = \frac{4\varepsilon \cos^2\left(2\pi\sqrt{\frac{E}{\hbar\Omega}}\right)}{\pi\left(-1 + \varepsilon + \sqrt{1 - 2\varepsilon}\,\cos\left(2\pi\sqrt{\frac{E}{\hbar\Omega}}\right)\right)}.
$$

Figure 8.5 shows the DOS for $\varepsilon \neq 0$. The levels increasingly broaden around the quantized energies of the decoupled ring ($\varepsilon = 0$) as ε increases. The uncoupled quantized values correspond to the poles of the density of states at zero coupling, which obey the relation $E = m^2\hbar\Omega$, with m an integer (values $m^2 = 0, 1, 4, 9$ in figure). When coupling is turned on, the levels are shifted to lower energies as the levels broaden, as expected in general for complex self-energy corrections. Deeper levels are less coupled to the reservoir than the shallower counterparts since there is partial transmission to the reservoir lead.

Making a correspondence between level broadening and electron lifetime, by fitting the resonance to a Lorentzian form leads to Figure 8.6. A power law decay of the lifetime with the reservoir coupling is observed for the smaller couplings.

The magnetic field shifts the states and the SO coupling breaks the twofold degeneracy as was discussed. Figure 8.8, depicts both the effect of the field and the SO. In panel a) each peak is doubly degenerate, while this degeneracy is broken with SO as depicted in panel b). Nevertheless this degeneracy can appear to exist when the coupling to the reservoir is sufficiently large (see panel b) for $\varepsilon = 0.5$).

8.7 PERSISTENT CHARGE CURRENTS

At zero temperature, the charge persistent currents in a decoupled ring can be calculated by the linear response relation [95, 52] $J_q = -\sum_i \frac{dE_i}{d\Phi}$ where i encloses the occupied states. The leading contribution to the current, due to cancellation of current contributions from state with opposite slopes, are the states close to the Fermi level. The linear response relation is not useful for the case we have coupling to the reservoir, since the energy broaden into a continuum of levels. On the other hand we have derived the exact wave

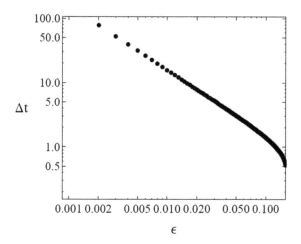

Figure 8.6: Lifetime of the electrons in the ring as a function of ε. The energies correspond to the quantized values of the decoupled states. The time is given in atomic units $1a.u. \approx 2.4 \times 10^{-17} s$.

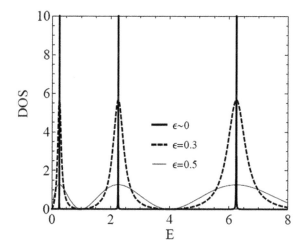

Figure 8.7: The density of states of the ring as a function of the energy for three values of ε and $T = 0$. The value of the parameters is $\omega_{SO}/\Omega = 0$, $\Phi/\Phi_0 = 0.5$.

functions from which the current may be determined by the expectation value of the charge current operator $\Psi^\dagger e v_\varphi \Psi$ where

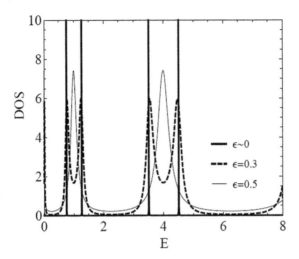

Figure 8.8: The density of states of the ring as a function of the energy for three values of ε and $T = 0$. The value of the parameters is $\omega_{SO}/\Omega = 0.75$, $\Phi/\Phi_0 = 0$.

$$v_\varphi = a\dot{\varphi} = (a/i\hbar)[\varphi, H] = -2a\Omega \left(-i\frac{\partial}{\partial\varphi} - \frac{\Phi}{\Phi_0} + \frac{\omega_{SO}}{2\Omega}\sigma_\rho \right), (8.54)$$

and integrating over all occupied states up to the Fermi level including the electron occupation numbers.

$$J_q = -\frac{2\varepsilon\hbar\Omega}{\Phi_0} \sum_{m,\mu} \int \frac{dE}{\hbar\Omega} \frac{f(E)}{\sqrt{\frac{E}{\hbar\Omega}} g^\mu} \sin^2(\pi n_m^\mu) \left[n\frac{\mu}{m} - \frac{\Phi}{\Phi_0} + \delta^\mu \right] (8.55)$$

with

$$\delta^1 = \sin^2\frac{\theta}{2} + \frac{\omega_{SO}}{2\Omega}\sin\theta; \quad \delta^2 = \cos^2\frac{\theta}{2} - \frac{\omega_{SO}}{2\Omega}\sin\theta, \qquad (8.56)$$

where $\overline{m}$ is the complement value of m and a natural current scale $J_0 = \hbar\Omega/\Phi_0$ is identified. Note that $\varepsilon = 0$ does not imply zero current (in fact it is largest at zero coupling) as g^μ also depends on the coupling with a nontrivial limit behavior. We will separate the discussion into two cases: I) The Fermi level fixes $N = 6$ electrons, (see Fig 8.9 top panel)) and II) $N = 8$, (see Fig 8.9 bottom panel)). In the absence of RSO interaction for the first case, there are two electrons, one with spin up and the other with spin down at each energy. At the Fermi level, two bands which

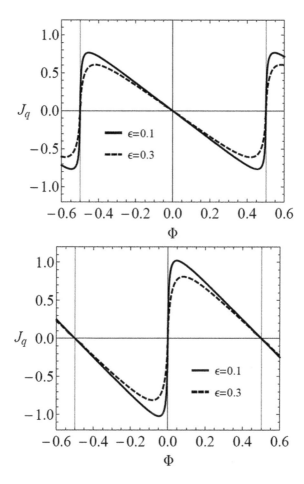

Figure 8.9: Charge persistent current as a function of the magnetic flux for three values of the reservoir coupling parameter. The number of electrons is 6 (top) and 8 (bottom) corresponding to a Fermi energy of $E_{F_1} =$ and $E_{F_2} =$. The RSO interaction is off and the persistent current is given in units $J_0 = \hbar\Omega/\Phi_0$.

describe electrons with different propagation numbers j cross each other at half integer steps in Φ_0 (see Figure **??**). This results in a jump in the sign of the current at these values. In the second case the levels cross at zero or integer flux quanta, and the sign jump occurs at those points. These are the behaviors expected also for small couplings to the reservoir. Figure 8.9 shows the charge currents without the SO coupling as a function of the magnetic field. The reduction in amplitude of the current as a function of the coupling strength is evident as decoherence increases. For Fermi level E_{f_1} the persistent current is minimal for the smallest fluxes and gradually grows, while for E_{f_2} the current is maximal at the smallest fluxes and decreases thereof.

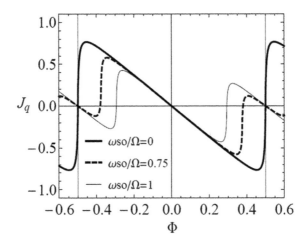

Figure 8.10: Charge persistent current as a function of the magnetic flux for different RSO values. The coupling is $\varepsilon = 0.1$ with the reservoir. The number of electrons is 6 (top) and 8 (bottom) corresponding to a Fermi energy of $E_{F_1} = 2.25$ and $E_{F_2} = 4$.

After including RSO, the crossing between bands at the Fermi level shift to $\Phi/\Phi_0 = m/2 + (1 \pm \sqrt{1 + (\omega_{SO}\Omega)^2})/2$, with $m \in \mathbb{Z}$ for the case I) and $\Phi/\Phi_0 = m/2 + \pm\sqrt{1 + (\omega_{SO}\Omega)^2}/2$ for case II) displacing the current jumps and introducing two more for each of the Fermi level scenarios, See Fig 8.10.

The degradation of current with temperature has a distinctive character as compared to the coupling to the reservoir, as can be seen in Figure 8.12. The temperature effect will be small when the current emanates from a level appreciably below the Fermi level, so that few electrons are actually promoted to counter current states. On the other hand, for fluxes where the current are from levels close to the Fermi level, the currents quickly degrade. So this is a mechanism that is energy dispersion dependent. Such mechanism is also observed in other scenarios, for instance graphene rings which will be the subject of Chapter 7. Figure 8.13 shows the dependence of charge current on temperature, for different ring-reservoir couplings. For certain ranges of the magnetic flux, the persistent current can be

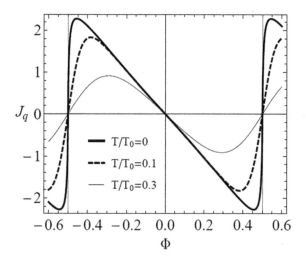

Figure 8.11: Charge persistent current as a function of the magnetic flux for different temperatures with $\varepsilon = 0.1$. The number of electrons is 6 (bottom) and 8 (top). Corresponds to $E_F = 2.25$.

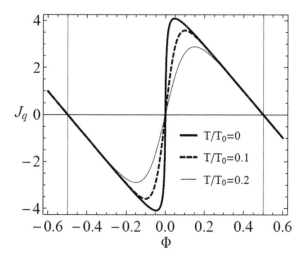

Figure 8.12: Charge persistent current as a function of the magnetic flux for different temperatures with $\varepsilon = 0.1$. The number of electrons is 6 (bottom) and 8 (top). Corresponds to $E_F = 4$.

degraded completely. We estimate the magnitude of the thermal effects by using the temperature scale $T_0 = \hbar\Omega/k_B$, As Ω depends on the size of the ring, $T/T_0 = 0.5$ in the figures, correspond to temperatures between

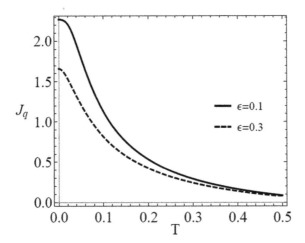

Figure 8.13: Temperature dependence of the charge current for 6 electrons. The maximal degradation is obtained with a combination of the coupling with the reservoir and increased temperature.

526 mK and 59 mK for ring sizes between 100 nm and 300 nm and an effective mass of $m^* = 0.042m_e$. This implies that the gap for persistent current degradation is of the order of $40\mu eV$ for the smallest of the rings. Improving this gap with either effective mass of ring radius and flux point of operation, might improve the thermal robustness of high sensitivity cantilevers for noise and electron thermometry.

8.8 PERSISTENT SPIN CURRENTS

The standard calculation is through the anticommutator of the velocity with the spin operator [52, 95, 102],

$$\vec{J}_s = \frac{\hbar}{4} \Psi^\dagger \{\vec{\sigma}, \vec{v}\} \Psi.$$

Appealing to the coupled ring wave function derived above and the velocity operator in Eq. (8.54) one can derive the spin current as, $J_s^z =$

$$-\varepsilon\hbar\Omega \sum_{m,\mu} \int \frac{dE}{\hbar\Omega} \frac{f(E)}{\pi\sqrt{\frac{E}{\hbar\Omega}}\, g^\mu} \sin^2(\pi n_m^\mu) \left[\left(n\frac{\mu}{m} - \frac{\Phi}{\Phi_0} \right) \beta^\mu + \gamma^\mu \right] (8.57)$$

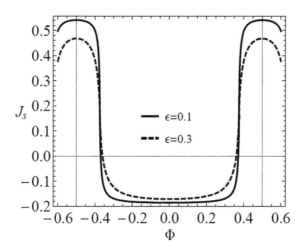

Figure 8.14: Spin persistent current as a function of the magnetic flux for three values of ε. The number of electrons is 6 (top) and 8 (bottom). The RSO is $\omega_{SO}/\Omega = 0.75$. The current is given in units of $J_0 = \hbar v_f/4\pi a$. Corresponds to $E_F = 2.25$.

with

$$\gamma^1 = \sin^2 \frac{\theta}{2}, \quad \gamma^2 = -\cos^2 \frac{\theta}{2},$$
$$\beta^1 = \cos \theta, \quad \beta^2 = 1.$$

In Figure 8.14 is drawn the spin persistent current as a function of the magnetic flux for the two Fermi levels considered in Figure 8.4. Spin currents are only possible in the presence of SO coupling, since spin degeneracy matches up identical contributions in charge current from opposite spins (see Figure 8.4 top panel). In the presence of the SO coupling there is a breaking of spin degeneracy with preservation of the time reversal symmetry, the necessary ingredients for their presence. As for charge currents, spin currents from deep levels in the Fermi sea, also tend to cancel but in a more complicated fashion. Figure 8.16 shows the combinations of charge currents with their corresponding spin orientations for the first Fermi level scenario: Deep in the Fermi sea charge currents are also paired up in spin but with small differences in electron velocities due to broken degeneracy. So we can see a small spin current accrued coming from theses levels. As one goes higher in magnetic field the positive current levels slow down, making less of a contribution, while the level with negative charge

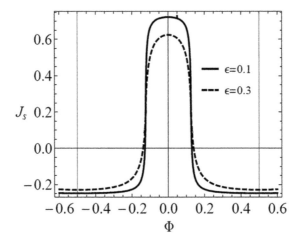

Figure 8.15: Spin persistent current as a function of the magnetic flux for three values of ε. The number of electrons is 6 (top) and 8 (bottom). The RSO is $\omega_{SO}/\Omega = 0.75$. The current is given in units of $J_0 = \hbar v_f/4\pi a$. Corresponds to $E_F = 4$.

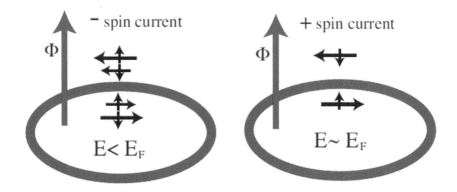

Figure 8.16: The figure depicts, qualitatively, the contributions to the spin current as the flux changes until the Fermi level is reached. On the left, the currents in each direction are highly compensated in spin (each current direction contains both spin directions). On the right, the flux is such that the energy is close to the Fermi level E_{F_1}, and the spin current is large and switches direction.

currents speed up, making the bulk of the current. The dispersion being quadratic makes for precise compensation, so that the full spin current is constant.

When the flux is large enough for the levels to cross the Fermi level, there is an abrupt disappearance of the negative spin up current and a new contribution from a positive spin up charge current, as shown in Figure 8.16 right panel. These two contributions make for a pure spin current, more than three times the magnitude of the previous regime, very close to the Fermi level E_{F_1}. The range of fluxes in which this happens is as wide as it takes for the second level to emerge from the Fermi sea, i.e., $\Delta(\Phi/\Phi_0) = \sqrt{1 + (\omega_{SO}/\Omega)^2} - 1$, at which point we start with the scenario on the left panel and repeat the whole periodic oscillation. For the second Fermi level the scenario is identical but it occurs for small fluxes in the center of the spectrum (Fig.8.14 bottom panel). Figure 8.19 shows how the spin currents, coming from different parts of the spectrum explored by the magnetic flux, can be tuned by the spin-orbit interaction at fixed coupling to the reservoir. One can see how positive and negative spin currents can be enhanced and change the range of fluxes for which they arise.

It is interesting to note that the smaller spin current coming from levels deeper in the Fermi sea is more robust to decoherence than the contributions coming from close to the Fermi level, resembling thermal effects previously discussed. On the other hand, as discussed for the charge currents, the Buttiker model is unable to completely degrade spin currents.

The coupling with the reservoir acts degrading the charge currents, consequently the spin current diminish its magnitude since the electrons lifetime on those states is lesser but not produces total decoherence. See Fig 8.17. The intensity of the spin current is also modified in this scenario by the SO interaction, the strength of the Rashba coupling, which can be modulated by an electric field perpendicular to the plane of the ring, acts on the local phases as Aharonov-Casher contribution, changing the polarization on the crossing points in the spectrum as can be seen in Figure 8.19.

8.9 EQUILIBRIUM CURRENTS IN A MESOSCOPIC GRAPHENE RING

As an interesting alternative to semiconductors ring we can focus on graphene as active media, with SO interaction, both due to its crystal field

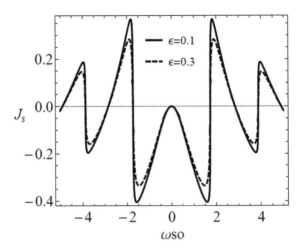

Figure 8.17: Spin persistent current as a function of the SO coupling for three values of ε. The number of electrons is 6 (top) and 8 (bottom). Corresponds to $E_F = 2.25$.

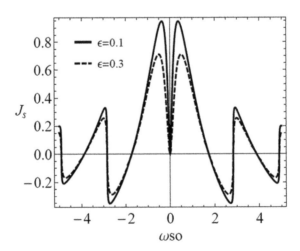

Figure 8.18: Spin persistent current as a function of the SO coupling for three values of ε. The number of electrons is 6 (top) and 8 (bottom). Corresponds to $E_F = 4$.

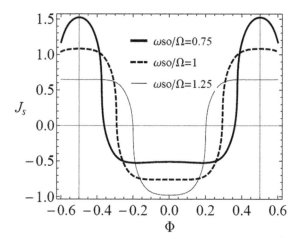

Figure 8.19: Spin persistent current as a function of the magnetic flux for three values of ω_{so}/Ω with finite temperature, $T/T_0 = 0.1$ and $\epsilon = 0.1$. The number of electrons is 6 (top) and 8 (bottom). Corresponds to $E_F = 2.25$.

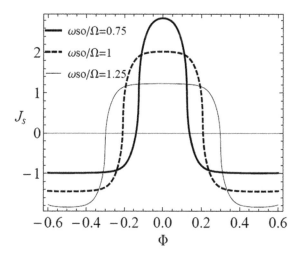

Figure 8.20: Spin persistent current as a function of the magnetic flux for three values of ω_{so}/Ω with finite temperature, $T/T_0 = 0.1$ and $\epsilon = 0.1$. The number of electrons is 6 (top) and 8 (bottom). Corresponds to $E_F = 4$.

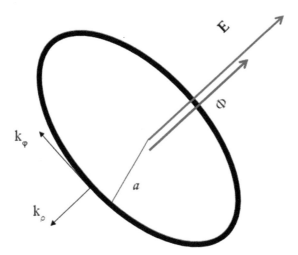

Figure 8.21: A quasi one-dimensional graphene ring in the long wavelength approach inherits all the internal symmetries the lattice structure. The Corbino ring is cut out from a graphene monolayer. The figure shows the (ρ, φ) coordinates, z being perpendicular to the plane. The average radius of the ring is a, represented by the red line. The ring is pierced by a perpendicular magnetic flux and electric field.

(ISO) and external field (Rashba type) coupling. A quasi one-dimensional ring of finite width, is cut out from a flat graphene monolayer as shown in Figure 8.21. Localized and discrete confined modes exist in the radial direction (see [54] for the case of carbon nano ribbons), while the angular direction is free, though appropriate closing conditions for the wave functions are to be applied. The Hamiltonian is given in[5], where there is SO interaction in the so called long wavelength limit around the Dirac points,

$$
\begin{aligned}
H = &-i\hbar v_F(\vec{\tau}_z\vec{\sigma}_x\mathcal{I}_s\partial_x + \mathcal{I}_\tau\vec{\sigma}_y\mathcal{I}_s\partial_y) \\
&+\Delta_{SO}\vec{\tau}_z\vec{\sigma}_z s_z + \lambda_R(\vec{\tau}_z\vec{\sigma}_x s_y - \mathcal{I}_\tau\vec{\sigma}_y s_x).
\end{aligned} \tag{8.58}
$$

The first term is the kinetic energy, it has the form $v_F\boldsymbol{\sigma}\cdot\mathbf{p}$, with the additional $\vec{\tau}_z$−Pauli matrix which acts on the "valley" index and distinguishes between the Dirac points in the band structure, $\mathbf{k} = \tau\mathbf{K} = \tau(4\pi/3c, 0)$ where c is the distance between the bravais lattice points and τ takes on values of ± 1. The $\vec{\sigma}_i$−Pauli matrices encodes for the sublattice distinction. The s_i represents the real spin of the charge carriers. Products of matrices in the Hamiltonian are understood as tensor products between different sub-spaces.

Where to clarify the usual notation, when only two operators or less are present, identity 2×2 matrices is implied for each of the omitted subspaces.

The second term in (8.58) is the intrinsic spin orbit (ISO) coupling[18], that due to the electric fields of the carbon atoms. From a tight binding point of view, this interaction comes from second neighbor hopping contribution that preserves all the symmetries of graphene. The last term in Hamiltonian (8.58) is the Rashba SO interaction which results from the action of an external electric field that breaks the spin mirror symmetry[55].

The operators that act on spaces σ_i and s_i are dimensionless and normalized as $\sigma_i^2 = \mathcal{I}_\sigma$ and $s_i^2 = \mathcal{I}_s$ (Pauli matrices), and the parameters Δ_{SO} and λ_R have dimensions of an energy.

Since the valley operator τ_z is diagonal, the Hamiltonian can be split into two different contributions, one for each valley in **k** space, thus reducing the model to two copies of a 4×4 matrix system, instead of the full 8×8 direct product space. One then gets two separate valley Hamiltonians,

$$H_+ = -i\hbar v_F(\vec{\sigma}_x \partial_x + \vec{\sigma}_y \partial_y)$$
$$+\Delta_{SO}\vec{\sigma}_z s_z + \lambda_R(\vec{\sigma}_x s_y - \vec{\sigma}_y s_x), \qquad (8.59)$$
$$H_- = -i\hbar v_F(-\vec{\sigma}_x \partial_x + \vec{\sigma}_y \partial_y)$$
$$-\Delta_{SO}\vec{\sigma}_z s_z - \lambda_R(\vec{\sigma}_x s_y + \vec{\sigma}_y s_x). \qquad (8.60)$$

so each valley can be treated separately.

8.10 RING HAMILTONIAN AND BOUNDARY CONDITIONS

A few important general points for the Hamiltonian in polar coordinates are worth to review, since they are frequently overlooked in the literature. The first that concerns us is the boundary conditions imposed on the wave function in a non-simply connected geometry, and the other, the form of the Hamiltonian used when a change in coordinate system is involved. We will first derive the closed ring Hamiltonian of radius a of pure kinetic energy by performing the proper coordinate mapping[52]. Then we will discuss the boundary conditions (BCs) on the ring geometry. The salient features of this model relevant to the full Corbino disk Hamiltonian will be transparent.

When using coordinates other than Cartesian, one must take care of subtleties in constructing an Hermitian Hamiltonian[56, 52], whose correct form avoids spurious features in the spectrum. In the $\tau = 1$ valley, keeping only the kinetic energy and omitting the spin degree of freedom, the coordinate change applied to Eq. (8.59) results in

$$H = -i\frac{\hbar v_F}{a}(\vec{\sigma}_y \cos\varphi - \vec{\sigma}_x \sin\varphi)\partial_\varphi, \tag{8.61}$$

after removing the radial part. The difficulty comes from the observation that this Hamiltonian is not Hermitian[57], since $\langle F \mid H \mid G \rangle^* \neq \langle G \mid H \mid F \rangle$, where $\mid F \rangle$ and $\mid G \rangle$ are 2-components spinors. This can be repaired by adding to Eq. (8.61) a term proportional to $i(\vec{\sigma}_y \sin\varphi + \tau\vec{\sigma}_x \cos\varphi)$ [52] and one is easily led to the form,

$$\begin{aligned} H_\tau = - \, & i\frac{\hbar v_F}{a}[(-\tau\vec{\sigma}_x \sin\varphi + \vec{\sigma}_y \cos\varphi)\partial_\varphi + \\ & - \tfrac{1}{2}(\vec{\sigma}_y \sin\varphi + \tau\vec{\sigma}_x \cos\varphi)]. \end{aligned} \tag{8.62}$$

Not including this term would also lead to real, but physically incorrect eigenvalues and eigenstates[57]. The reason for real eigenvalues, in spite of non-hermiticity, follows from the operator being PT (parity and time reversal) symmetric[58].

As in [52] the derivation of the correct Hamiltonian can be pictured as follows: Le us start with the Hamiltonian $H = v_F\boldsymbol{\sigma} \cdot \mathbf{p}$, changing directly in polar coordinates, one gets

$$H = -i\hbar v_F\left(\vec{\sigma}_\rho\partial_\rho + \rho^{-1}\vec{\sigma}_\varphi\partial_\varphi\right), \tag{8.63}$$

fixing $\rho = a$ and taking care to properly symmetrize the product $\vec{\sigma}_\varphi\partial_\varphi$, one obtains a ring

$$H_{\text{ring}} = -i\hbar v_F a^{-1}(\vec{\sigma}_\varphi\partial_\varphi - \tfrac{1}{2}\vec{\sigma}_\rho), \tag{8.64}$$

that corresponds to the expression in (8.62) using $\vec{\sigma}_\rho = \vec{\sigma}_x \cos\varphi + \vec{\sigma}_y \sin\varphi$ and $\vec{\sigma}_\varphi = -\vec{\sigma}_x \sin\varphi + \vec{\sigma}_y \cos\varphi$. The term $\tfrac{1}{2}\vec{\sigma}_\rho$ in (8.64) is essential since it renders the derivative in polar coordinates covariant by introducing the connection that correctly rotates the internal degree of freedom so as to keep the pseudo spin parallel to the momentum. This form of the Hamiltonian for the angular dependence is arrived independently in the shape of the confining potential applied radially[56, 54]. The details of the confining potential will arise as effective coefficients in the form of (8.64).

The eigenstates of (8.64) are of the form

$$\psi(\varphi, z) = \frac{e^{im\varphi}}{\sqrt{2}} \begin{pmatrix} -i\kappa e^{-i\varphi} \\ 1 \end{pmatrix}, \tag{8.65}$$

with m a half positive integer for metallic rings and $\kappa = \pm 1$ describing electrons and holes respectively. The corresponding energies are $E = \frac{\kappa\hbar v_F}{a}(m-1/2)$. It is easy to verify that $\langle\boldsymbol{\sigma}\rangle = \kappa(-\sin\varphi, \cos\varphi) = \kappa\mathbf{k}/|\mathbf{k}|$, as can be derived in Cartesian coordinates. In spite of the fact that pseudo-spin follows the momentum, it is endowed with proper angular momentum[59]. This can be verified by noting that the Hamiltonian of (8.64) does not commute with the orbital angular momentum alone $L_z = -i\hbar\partial_\varphi$ but with the combination $J_z = L_z + \frac{1}{2}\hbar\sigma_z$ (and with $(L_z\mathbf{u}_z + \frac{1}{2}\hbar\boldsymbol{\sigma})^2$. Note that if L_z does not commute with the Hamiltonian, there is a torque on the orbital momentum. This torque is compensated by a torque on the pseudo spin angular moment so that the total J_z is conserved. Thus, with the pseudo spin there is associated "lattice spin" presumably from the rotation the electron sees of the A-B bond. We will see in the next section, how this extra angular momentum combines with the regular electron spin to generate a total conserved angular momentum.

The wave function nevertheless, preserves spin-like properties. One can verify that the ring eigenfunctions are anti-periodic, thus $\psi(\varphi + 2\pi) = -\psi(\varphi)$, a property which finds its origin in the effect of the 2π rotation on the connection $\frac{1}{2}\sigma_\rho$. The factor $\frac{1}{2}$ corresponds to a Berry phase, discussed as a very crucial feature of graphene (e.g. by Katsnelson [6], Guinea et al [16]) and of carbon nanotubes (e.g. in Ref. [60]). We will reemphasize these points in our derivation in the following sections which several previous references have overlooked (see references [57, 17, 61, 52]).

8.11   CLOSING THE WAVE FUNCTION ON A GRAPHENE RING

In this section, we are interested in discussing graphene rings described with the effective Dirac theory in the vicinity of the K points with appropriate boundary conditions. Recalling that according to Bloch's theorem, the wave function $\psi(\mathbf{r}) = u_{\mathbf{k}}(\mathbf{r})e^{i\mathbf{k}\cdot\mathbf{r}}$ should exhibit the ring periodicity, while the Bloch amplitude $u_{\mathbf{k}}(\mathbf{r})$ has the lattice periodicity. The periodic boundary conditions imposed on the Bloch wave function do not necessarily imply periodic boundary conditions for the eigenfunctions of the effective theory [18]. Indeed, $\mathbf{k}$ is measured from the Brillouin zone center (Γ point). The effective theory is related to the wave vector $\mathbf{q} = \mathbf{p}/\hbar$ in the neighborhood of the Dirac points through $\mathbf{k} = \mathbf{K}_D + \mathbf{q}$.

Generalizing the boundary conditions for the case of a ring with linear dispersion (see previous section) we introduce a twist phase θ_0 in the closing of the wave function

$$\psi(\varphi + 2\pi) = e^{-i\theta_0}\psi(\varphi). \tag{8.66}$$

The eigenstates are now of the form

$$\psi(\varphi) = e^{i(m-\theta_0/2\pi)\varphi}\begin{pmatrix} Ae^{-i\varphi} \\ B \end{pmatrix}, \tag{8.67}$$

with m an integer, with corresponding eigenvalues

$$E = \pm\frac{\hbar v_F}{a}\left|m - \frac{\theta_0}{2\pi}\right| \tag{8.68}$$

where $\kappa = \pm 1$ refers to particles (conduction band) and holes (valence band). As we discussed in the previous section, in the case of a graphene ring, antiperiodic BCs (ABC) should be chosen[18, 16]. This means $\theta_0 = \pi$ for graphene in a Corbino geometry, but for different boundary conditions, such as those that occur in carbon nanotubes with arbitrary chiralities, can also be described. Note that the twist phase plays the same role as a magnetic flux through the ring that can modify its conducting properties by manipulating the gap at the Dirac point.

Let us extend the boundary conditions discussion for the proposed ring. The discussion on graphene nano ribbons have been addressed in detail[72]. For the approximation addressed here, the zig-zag nano-ribbons are the closest relative, since it has been shown[19] that a generically cut honeycomb lattice has approximately zig-zag boundary conditions to a high accuracy. For graphene ribbons with zigzag edges there is the concern that longitudinal and transverse states are coupled[72] and slicing the graphene band using the boundary conditions is not warranted for small ribbon widths $N \sim 1$ (number of transverse lattice sites). Nevertheless, for wide ribbons ($N \gg 1$) this approximation becomes increasingly good as can be judged from the relation coupling the longitudinal k and transverse p modes $\sin pN + w\cos(k/2)\sin p(N+1) = 0$ where the wavectors in units of the magnitude of the primitive translation vectors of the lattice. When $N \gg 1$ then $p = m\pi/N$ independent of k. One final concern is the existence of one localized state that for nano ribbons for a critical value of the longitudinal wave vector, nevertheless, the restriction also disappears in the limit $N \gg 1$ in which our continuum approximation is based. In the next section we will discuss the possible coupling of the transverse modes due to the spin-orbit interaction.

The vicinity to the Dirac points is an important issue here, since the linear range of the spectrum is subject to the lattice parameter and the radius of the ring. The estimated limiting value of the momentum ignoring lattice effects[73] is $k_l \approx 0.25\text{nm}^{-1}$. The carrier limiting energy at this point is $E_l = \hbar v_F k_l$. Equating this value with (8.68) we obtain the maximum number of states, hence,

$$\left| m - \frac{\theta_0}{2\pi} \right| \lesssim k_l a. \qquad (8.69)$$

As a reference estimation based on an analogous ring already present in nature (in fact a carbon nanotube section has a kinetic term of the same form as the Corbino). A single wall carbon nanotube has radius that goes from 10 nm to 100 nm, this gives order of magnitudes from $m \sim 2$ to $m \sim 25$ that varies depending whether it is an armchair or zigzag tube. For a carbon nanotube with a smaller radius than 4 nm, the allowed states will be outside the linear region establishing a threshold for the values of a in the long wavelength approach.

8.12 SPIN-ORBIT COUPLING

Having set up the correct Hamiltonian and boundary conditions to describe a graphene ring, we can incorporate SO interactions in the ring geometry to obtain the equivalent of Eqs. (8.59) and (8.60). The spectrum becomes independent of the valley index τ, so we will only deal with $\tau = +1$ in polar coordinates:

$$H_+ = -i\hbar v_F a^{-1}(\vec{\sigma}_\varphi \partial_\varphi - \tfrac{1}{2}\vec{\sigma}_\rho)$$
$$+\Delta_{SO}\vec{\sigma}_z s_z + \lambda_R(\vec{\sigma}_\rho s_\varphi - \vec{\sigma}_\varphi s_\rho). \tag{8.70}$$

For the ISO only case we assume a 4−component vector to represent the electronic states, incorporating electron spin, $\Psi = e^{im\varphi}\left(A_\uparrow^{\kappa,\delta}e^{-i\varphi}, A_\downarrow^{\kappa,\delta}, B_\uparrow^{\kappa,\delta}, B_\downarrow^{\kappa,\delta}e^{i\varphi}\right)^T$, where $A_{\uparrow,\downarrow}^{\kappa,\delta}$ $(B_{\uparrow,\downarrow}^{\kappa,\delta})$ is the wave vector amplitude on sublattice A (B) with spin $\delta = \uparrow\downarrow$. The ansatz for the spinor is constructed in order to account for the conservation of the total angular momentum. All components carry the same angular momentum J_z, adding in units of $\hbar$ a purely orbital contribution (respectively $m - 1$, m, m, and $m + 1$ for the four components), a pseudo-spin or lattice contribution (resp. $+\tfrac{1}{2}, +\tfrac{1}{2}, -\tfrac{1}{2}$ and $-\tfrac{1}{2}$), and the spin contribution (resp. $+\tfrac{1}{2}$, $-\tfrac{1}{2}, +\tfrac{1}{2}$ and $-\tfrac{1}{2}$).

The eigenenergies, assuming these wave functions (with constant amplitudes A an B), are

$$E_{m,\Delta}^{\kappa,\delta} = \kappa\sqrt{\Delta_{SO}^2 + \epsilon^2(m - \delta/2)^2}, \tag{8.71}$$

where $\kappa = \pm 1$ is the particle-hole index and $\delta = \pm 1$ the SO index, and $\epsilon = \frac{\hbar v_F}{a}$. $\kappa\epsilon|(m - \delta/2)|$ corresponds to the electron energies in the absence of ISO. On the other hand when only the Rashba interaction is present the energy is given by

$$E_{m,\lambda_R}^{\kappa,\delta} = \frac{\kappa}{2}\sqrt{\epsilon^2(1 + 4m^2) + 8\lambda_R^2 - 4\delta f}$$
$$where f = \sqrt{(m^2\epsilon^2 + \lambda_R^2)(\epsilon^2 + 4\lambda_R^2)}, \tag{8.72}$$

which has the correct zero SO coupling limit. These energies correspond to the angular wave vectors satisfying the closed ring boundary conditions.

The spectrum is shown in Figure 8.22. We assume that the transverse mode is in the basis state using again as reference the transverse modes for the graphene zigzag ribbons. The spinor wave functions for the ribbons depend on both longitudinal and transverse indexes. Choosing the basis state in the $N \gg 1$ limit permits writing an explicit expression for the wave functions and assess the coupling of the free transverse modes in the presence of the SO couplings. If the coupling is large compared to the transverse level separation, it must be contemplated in our analysis[74]. Writing the ribbon wave functions in the basis above, we computed that in the $N \gg 1$ limit the coupling between transverse modes is negligible. So that the single transverse mode approximation is warranted.

Although the possible wave vectors take on discrete half integer values, they will trace a continuous change when a gauge field is applied. Close to the point of closest approach between the valence and conduction bands. For the ISO coupling these points are around $m = \pm 1/2$ and the expansion takes the form

$$E_{m,\Delta}^{\kappa,\delta} = \kappa|\Delta_{SO}| + \frac{\kappa\epsilon^2}{2|\Delta_{SO}|}(m \pm 1/2)^2 + O\left((m \pm 1/2)^4\right), \quad (8.73)$$

while for the Rashba coupling the behavior is

$$E_{m,\lambda_R}^{\kappa,\delta} = \frac{\kappa|m \pm 1/2|}{\sqrt{2(\epsilon^2 + 4\lambda_R^2)}} + O\left((m \pm 1/2)^2\right).$$

$$(8.74)$$

The intrinsic spin-orbit term will open a gap in the vicinity of $(m = \pm 1/2)$ which is simply $2\Delta_{SO}$ where the electrons exhibit an effective mass of $m_{ISO}^* = \Delta_{SO}^2/v_F^2$ which is small, both because v_F is large and Δ_{SO} is in the range of meV for graphene. For the Rashba coupling there is no gap at $m = \pm 1/2$ but we will see a spin dependent gap opens continuously as the magnetic field is applied. Note also that this is a gap between spin-orbit up states. The gap between spin-orbit down states is given by $\sqrt{\epsilon^2 + 4\lambda_R^2} + 2\lambda_R$. One can define an effective mass of the spin down states as $m_\downarrow^* = \kappa\lambda_R\hbar^2/[2\epsilon^2(2\lambda_R + \sqrt{\epsilon^2 + 4\lambda_R^2})]$.

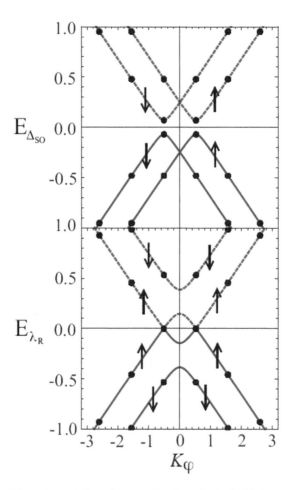

Figure 8.22: Dispersion relations for metallic rings for both SO interactions (intrinsic middle and Rashba bottom, panels). The ISO has been drawn for $\Delta_{SO} = 0.25$, and opens a gap of size $2\Delta_{SO}$ with separate branches for each spin label. The Rashba interaction is depicted for $\lambda_R = 0.4$, the allowed values of m are indicated by the full dots (Note: the spin asymmetry introduced by the Rashba coupling, that will have striking consequences for the charge and spin persistent currents).

The limit in which the SO coupling goes to zero is singular, since both gaps close and the dispersion becomes linear as $\kappa \hbar v_F q_\varphi$. This limit highlights another feature of the Rashba spectrum. In the vicinity of the Dirac points K, and K', the electron behaves as a hole (has negative mass) in the conduction band and has negative charge (positive mass). From the expression above $m_\uparrow^* = -m_\downarrow^*$ at the Dirac point.

The typical values used for intrinsic coupling are estimated in Ref. [55] using a microscopic tight-binding model with atomic spin orbit interaction. The Rashba interaction comes from the atomic spin orbit and Stark interactions and the intrinsic from the mixing between σ and π bands due to atomic spin orbit interaction. The coupling constants are given by the expressions,

$$\Delta_{SO} = \frac{|s|\xi^2}{18(sp\sigma)^2},$$

$$\lambda_R = \frac{eEz_0\xi}{3(sp\sigma)},$$

where $|s|$ and $(sp\sigma)$ are hopping parameters in the tight-binding model, $s = -8.868$ eV and $(sp\sigma) = 5.580$ eV, $\xi = 6$ meV is the atomic SO strength of carbon, and $z_0 \sim 3 \times a_B$ (a_B is the Bohr radius), is proportional to its atomic size. λ_R is proportional to the electric field, $E \approx 50$ V/300 nm, perpendicular to the graphene sheet. This gives values for the SO parameters $\lambda_R \approx 0.1$ K and $\Delta_{SO} \approx 0.01$ K.

The split bands open a gap symmetrically between the δ states when $\Delta_{SO} = 0$. If $\Delta_{SO} \neq 0$ the contributions for each gap are different [75, 76]. In this parameterization the blue and the red curves (dashed and continuous respectively) represent the levels in the quantization axis of the RSO interaction, i.e., in the SO basis[49].

As we will see below, the velocity operator merits a non-trivial treatment in the context of graphene. For this reason we will derive the eigenfunctions for both SO couplings to compute the charge and spin persistent currents using the velocity operator, and compare it with the linear response relation. For the ISO only we have the wave functions

$$\Psi_{m,\Delta}^{\kappa,\delta}(\varphi) = \frac{e^{im\varphi}}{2|E_{m,\Delta}^{\kappa,\delta}|} \begin{pmatrix} \delta_{\delta,+}[\epsilon(m-\tfrac{1}{2})-i(\Delta_{SO}+E_{m,\Delta}^{\kappa,\delta})]e^{-i\varphi} \\ \delta_{\delta,-}[\epsilon(m+\tfrac{1}{2})+i(\Delta_{SO}-E_{m,\Delta}^{\kappa,\delta})] \\ \delta_{\delta,+}[\epsilon(m-\tfrac{1}{2})-i(\Delta_{SO}-E_{m,\Delta}^{\kappa,\delta})] \\ \delta_{\delta,+}[\epsilon(m+\tfrac{1}{2})+i(\Delta_{SO}+E_{m,\Delta}^{\kappa,\delta})]e^{i\varphi} \end{pmatrix},$$

$$(8.75)$$

labeled by κ and δ as $\Psi^{\kappa,\delta}_{m,\Delta}$. The polarization of this state is given by the expectation value of the operator $(\hbar/2)\mathcal{I}_\sigma \vec{s}$,

$$\langle s_z \rangle = \frac{\hbar}{2}(\Psi^{\kappa,\delta}_{m,\Delta}(\varphi))^\dagger \mathcal{I}_\sigma s_z \Psi^{\kappa,\delta}_{m,\Delta}(\varphi). \tag{8.76}$$

and all the states are polarized perpendicular to the Corbino disk, i.e., the $\mathbf{z}$ direction. This is also the direction of the effective magnetic field implied by the rewriting of the ISO term as $(\Delta_{SO}\boldsymbol{\sigma}) \cdot \vec{s} = (\Delta_{SO}\sigma_z)s_z$, a field that aligns the spins in opposite direction on different sublattices, in the z direction. The result is zero global spin-magnetization while each sub lattice is spin-magnetized in opposite directions. This is in accordance with the fact that the intrinsic SO interaction operates as a local magnetic field in each sublattice with opposite sign, and thus not breaking of time reversal symmetry.

The pseudo spin polarizations are computed in an analogous fashion

$$\langle \boldsymbol{\sigma} \rangle = \frac{\hbar}{2}(\Psi^{\kappa,\delta}_{m,\Delta}(\varphi))^\dagger \boldsymbol{\sigma}\mathcal{I}_s \Psi^{\kappa,\delta}_{m,\Delta}(\varphi),$$
$$= \frac{\kappa\hbar}{2}\frac{\delta\hat{z}\Delta_{SO} + (m - \delta/2)\epsilon\hat{\varphi}}{E^{\kappa,\delta}_{m,\Delta}}, \tag{8.77}$$

where we note the ordering go the matrix direct product. One sees both orbital and spin-orbit contributions, so the pseudo spin does not simply follow the electron momentum.

The Rashba eigenfunctions are

$$\Psi^{\kappa,\delta}_{m,\lambda_R}(\varphi) = \frac{e^{im\varphi}}{\sqrt{\Lambda}}\begin{pmatrix} \frac{-2iE^{\kappa,\delta}_{m,\lambda_R}(m\epsilon^2+2\lambda_R^2+\delta\Gamma_m)}{(4m^2-1)\epsilon^2\lambda_R}e^{-i\varphi} \\ \frac{-2iE^{\kappa,\delta}_{m,\lambda_R}}{\epsilon(2m+1)} \\ \frac{m\epsilon^2-2\lambda^2+\delta\Gamma_m}{\epsilon\lambda_R(2m+1)} \\ e^{i\varphi} \end{pmatrix},$$

$$\tag{8.78}$$

where $\Gamma_m = \sqrt{(m^2\epsilon^2 + \lambda_R^2)(\epsilon^2 + 4\lambda_R^2)}$ and $\Lambda = 4\Gamma_m\left(\Gamma_m - \delta(2\lambda_R^2 - m\epsilon^2)\right)/(2m + 1)^2\epsilon^2\lambda_R^2$. The polarization of the

Rashba eigenvectors is given by

$$\langle \vec{s} \rangle = \frac{\hbar}{2} (\Psi_{m,\lambda_R}^{\kappa,\delta}(\varphi))^{\dagger} \mathcal{I}_\sigma \vec{s} \Psi_{m,\lambda_R}^{\kappa,\delta}(\varphi),$$

$$= \delta \left(\frac{\hbar}{2} \right) \frac{m\epsilon(2\lambda_R \hat{\rho} + \epsilon \hat{z})}{\Gamma_m}, \tag{8.79}$$

where two contributions are evident, the polarization points outward in the radial direction and has a component due to the orbital rotation of the electrons.

Following previous expressions the Rashba pseudo-spin polarizations are

$$\langle \boldsymbol{\sigma} \rangle = \frac{\hbar}{2} (\Psi_{m,\lambda_R}^{\kappa,\delta}(\varphi))^{\dagger} \boldsymbol{\sigma} \mathcal{I}_s \Psi_{m,\lambda_R}^{\kappa,\delta}(\varphi),$$

$$= \frac{\hbar}{2} \frac{\delta m \epsilon E_{m,\lambda_R}^{\kappa,\delta}(\delta m \gamma + \Gamma_m)\hat{\varphi}}{(m - 1/2)(\delta m \epsilon^2 \Gamma_m + m^2 \epsilon^2 \gamma + \lambda_R^2(\gamma - 2\delta \Gamma_m))}, \tag{8.80}$$

where $\gamma = \epsilon^2 + 4\lambda_R^2$.

8.13 CHARGE PERSISTENT CURRENTS

Persistent equilibrium currents are a direct probe of energy spectrum of the system in the vicinity of the Fermi energy. Although such currents are typically small and are detected by the magnetic moment they produce[77], recent experiments, where many rings form dense arrays on a cantilever, boost the magnetic signal allowing both measurement of the current signal and the use of the set up as a sensitive magnetometer. The Corbino disk geometry can be easily built with high precision by using new techniques[78] manipulating nano-particles as cutters and hydrogenating the open bonds.

The spectrum of the system is modified by a field flux perpendicular to the Corbino disk as follows,

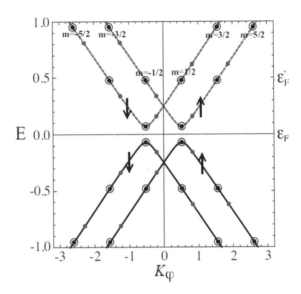

Figure 8.23: The energy dispersion for the ISO with $\Delta_{SO} = 7 \times 10^{-7}$ a.u. as a function of wave vector K_φ. The circled dots represent the allowed values of the energies on the ring at zero magnetic flux. The blue dots represent the shift of the allowed energies due to a finite flux.

$$E^{\kappa,s}_{m,\Delta}(\Phi) = \kappa\sqrt{\Delta^2_{SO} + \epsilon^2(m - \delta/2 + \Phi/\Phi_0)^2}, \tag{8.81}$$

$$E^{\kappa,\delta}_{m,\lambda_R}(\Phi) = \frac{\kappa}{2}\left(8\lambda^2_R + \epsilon^2\left(4(m + \Phi/\Phi_0)^2 + 1\right) - \right.$$
$$\left. -4\delta\sqrt{(4\lambda^2_R + \epsilon^2)\left(\lambda^2_R + \epsilon^2(m + \Phi/\Phi_0)^2\right)}\right)^{1/2} \tag{8.82}$$

where the Zeeman coupling has been neglected at small enough fields. The addition of a magnetic field, in the form of a $U(1)$ minimal coupling with flux Φ threading the ring, breaks time reversal symmetry allowing for persistent charge currents[9]. In the case of a ring of constant radius threaded by a perpendicular magnetic flux, the angular component of the gauge vector $A_\varphi = \Phi/2\pi a$ may be eliminated via a gauge transformation $A'_\varphi = A_\varphi + a^{-1}\partial_\varphi\chi = 0$, $\Psi'(\varphi) = \Psi(\varphi)e^{ie\chi/\hbar}$ at the expense of

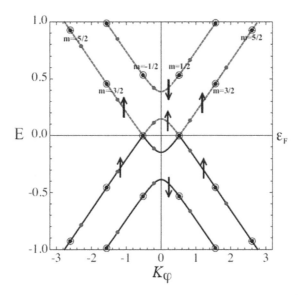

Figure 8.24: The energy dispersion Rashba coupling as a function of wave vector K_φ. The circled dots represent the allowed values of the energies on the ring at zero magnetic flux. The blue dots represent the shift of the allowed energies due to a finite flux. The Fermi energy is assumed to be zero, except for the Rashba where a finite value for the Fermi energy is also illustrated.

modifying the BCs on the ring to

$$\Psi'(\varphi + 2\pi) = e^{-i\theta_0} e^{-2i\pi\Phi/\Phi_0} \Psi'(\varphi), \qquad (8.83)$$

where Φ_0 is the normal quantum of magnetic flux (h/e). As mentioned before, the twist in the BCs and the field accomplish the same effect, so one can use them interchangeably while satisfying the relation

$$E^{\kappa,\delta}_{m,\Phi}(\theta_0) = E^{\kappa,\delta}_{m,0}(\theta_0 + 2\pi\Phi/\Phi_0), \qquad (8.84)$$

hence $m \to m - \theta_0/2\pi + \Phi/\Phi_0$, as discussed in Eq.(8.68). The energy dispersion for the graphene ring is illustrated in Figure 8.24 (left panel), where the different colors refer to the conduction band ($\kappa = +1$, dashed line) and valence band ($\kappa = -1$, full line). As expected, the energy levels display a periodic variation with the magnetic flux (right panel in the figure).

The charge persistent current in the ground state can be derived using the linear response definition $J_Q = -\sum'_{m,\kappa,\delta} \frac{\partial E}{\partial \Phi}$, where the primed sum refers to all occupied states only. Since the current is periodic in Φ/Φ_0 with a period of 1, we can restrict the discussion to the window $0 \leq \Phi < \Phi_0$ where the occupied states are in the valence band $\kappa = -1$, since the Fermi level is chosen at the zero of energy. We will first discuss the simple ISO coupling. The analytical expression is given by

$$J_{Q,\Delta}^{\kappa} = -\frac{\epsilon^2 \kappa}{\Phi_0} \sum_{m,\delta}' \frac{(m - \delta/2 + \Phi/\Phi_0)}{E_{m,\Delta_{SO}}^{\kappa,\delta}(\Phi)}. \tag{8.85}$$

In Figure 8.24, on the left panel, the spin-orbit branches of the spectrum labeled with their spin quantum number have been depicted. The encircled dots are the allowed energy values, due to quantization on the ring, at zero magnetic field. When the field is turned on, these dots are displaced (no longer encircled) on the energy curve.

On the right panel we depict the trajectory of these dots as the magnetic field is increased for both the filled (full lines in figure) and unfilled (dashed lines) states. The negative derivative of the curves on the right panel added over the occupied states (both spin quantum numbers) is the net charge persistent current. For the range of energies shown, the only net contribution is from the levels closest and below the Fermi level. The lower levels have currents that tend to compensate in pairs. Following the curve on the right, below the Fermi energy and from zero field, the current first increases linearly and then bends over to reach a maximum value before two levels cross (crossing indicated by arrow on the right panel of Figure 8.24). At that point, one follows the level closest to the Fermi energy (from below), the current changes sign and increases crossing the zero current level, whereupon the whole process repeats periodically. Such behavior is shown in Figure 8.26 top panel. Changing the Fermi level can change the scenario qualitatively. For example, adjusting the Fermi level to ε_F' (see Figure 8.24), the currents would follow a square waveform, alternating between constant current blocks of opposite signs.

For the Rashba coupling, represented in the bottom panels in Figure 8.24, the current is derived in a similar way, but now there is a striking asymmetry between spin branches. The analytical form for the charge

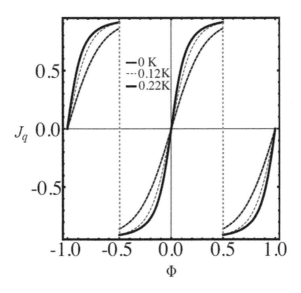

Figure 8.25: The equilibrium charge currents for both ISO interactions. The ring is considered to have a radius of $a = 20nm$, $\Delta_{SO} = 0.4$.

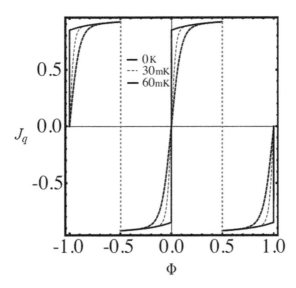

Figure 8.26: The equilibrium charge currents for Rashba interactions. The ring is considered to have a radius of $a = 20nm$, $\lambda_R = 0.25$.

current is

$$
J_{Q,\lambda_R}^{\kappa} =
$$

$$
-\frac{\epsilon^2 \kappa}{\Phi_0} \sideset{}{'}\sum_{m,\delta} \frac{\left(2 - \dfrac{\delta(\epsilon^2 + 4\lambda_R^2)}{\sqrt{(\epsilon^2 + 4\lambda_R^2)(\lambda_R^2 + \epsilon^2(m + \Phi/\Phi_0)^2)}} \right) (m + \Phi/\Phi_0)}{E_{m,\lambda_R}^{\kappa,\delta}(\Phi)}.
$$

$$(8.86)$$

The spin branch closest to the Fermi energy is non-monotonous, making for two different contributions to the charge current for the up spin contribution. Note also that we have taken into account the current coming from the spin down branch which does not have the same effective mass as the corresponding branch of the opposite spin. The results are depicted in Figure 8.26 bottom panel. The structure of the spectrum being asymmetric between spin branches makes for the possibility of net spin currents as we will see below.

The charge persistent current can be manipulated with λ_R since the Rashba parameter can be tuned by a field perpendicular to the plane of the ring. In contrast, the intrinsic SO cannot be easily tuned by applying external fields. Nevertheless, it has been established experimentally[79] that light covering of graphene with covalently bonded hydrogen atoms modifies the carbon hybridization and can enhance the intrinsic spin-orbit strength by three orders of magnitude[79]. Regulating this covering may then be a tool to manipulate charge currents.

One can contemplate the effect of temperature on the robustness of persistent charge currents by considering the occupation of the energy levels. The Fermi function has then to be factored into the computation of the currents

$$
J_{Q,\lambda_R}^{\kappa}(T) = -\sum_{m,\delta} \frac{\partial E_{m,\lambda_R}^{\kappa,\delta}(\Phi)}{\partial \Phi} f(E_{m,\lambda_R}^{\kappa,\delta}, \varepsilon_F, T), \qquad (8.87)
$$

where $f(E, \varepsilon_F, T) = (1 + \exp{(E - \varepsilon_F)}/k_B T)^{-1}$ is the Fermi occupation function for the case of the Rashba coupling. There is no need now to restrict the energy levels contemplated since the filling is determined by the Fermi distribution.

Figure 8.26, shows the effect of a temperature energy scale of the order of the SO strength for both intrinsic and Rashba couplings. The deep levels will be fully occupied while the shallow levels (close to the Fermi energy) will have a temperature dependent occupancy. Occupation depletion affects mostly the current contributions from levels within $k_B T$ of the Fermi level. This typically happens in the vicinity of the integer values of the normalized flux Φ/Φ_0, but at half integer fluxes the contributing levels dig into the Fermi sea where carrier depletion is less pronounced and current discontinuities tend to be protected from temperature effects. From Figure 8.24 one can estimate the depth in energy of the crossing to be $\sim 3 \times 10^{-4}$ a.u which amounts to a temperature equivalent of ~ 1 K before degradation of spin currents is observed at half integer fluxes. This is an important feature of the linear dispersions in graphene, and in enhanced SO coupling scenarios could be of applicability for magnetometer devices at relatively higher temperatures.

8.14 EQUILIBRIUM SPIN CURRENTS

We now contemplate spin equilibrium currents. In the absence of a direct linear response definition one can obtain them from the charge currents by distinguishing the velocities of different spin branches. We define a spin equilibrium current as

$$J_S = J_Q(\delta = -1) - J_Q(\delta = 1), \tag{8.88}$$

where one weighs the asymmetry in velocities of the different occupied spin branches. As we mentioned in the previous section there is no spin asymmetry both for the free case and for the ISO, so no spin current can result in this case, i.e., both spin branches contribute charge current with the same amplitude so they cancel in the above expression. With the Rashba coupling, the inversion symmetry is broken inside the plane and the spin branches are asymmetrical for a range of q_φ values.

The peculiar separation of the spin branches makes for velocity differences of the two spin projections and a spin current ensues as shown in Figure 8.27. The figure shows a large spin current for small fluxes that can be traced back to the large charge currents coming from a single spin

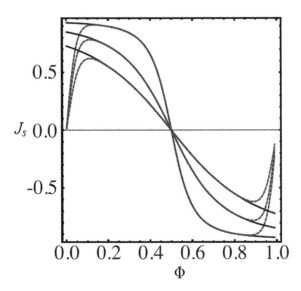

Figure 8.27: Spin current for the Rashba SO coupling indicated by the legends as a function of the magnetic field, as derived form the charge currents distinguished by spin components. The large spin currents at small fields are due to dominant charge currents for a single spin orientation. Temperature affect currents at the integer values of flux, while toward half integer values currents are protected.

branch in Figure 8.24. Toward half integer flux quantum's the opposite spin charge current increases until it cancels out the spin current completely. Beyond half integer flux the spin current is reversed in sign and at zero temperature there is a discontinuity approaching integer fluxes. As discussed for charge currents, the spin currents are also most susceptible to thermal depletion of carriers at integer fluxes, while toward half integer fluxes these are protected.

A striking feature, that survives temperature effects, is that the spin currents increase as one lowers λ_R. The Rashba coupling breaks inversion symmetry in the plane even for small λ_R. The symmetry breaking determines the spin labeling of the energy branches that take part in the spin current. It is only for $\lambda_R = 0$ that the free Hamiltonian symmetry is re-established and the spin currents are destroyed. A combination of the described symmetry effect and the thermal shielding from deep levels make for these effects observable experimentally.

8.15 VELOCITY OPERATORS FOR GRAPHENE

As discussed in Section 8.3, there are two ways to compute the effect of the magnetic field: either putting the description in the Hamiltonian as a gauge vector or performing a gauge transformation and passing all field information to the wave function. For $SU(2)$ gauge theory applied to the present case, this process cannot be done directly because of the lack of gauge symmetry[13]. We have solved the problem fully for the "gauge fields" in the Hamiltonian and determined the eingenfunctions. Such eigenfunctions contain the full information of the state, and the velocities as a function of the magnetic field can be derived by using the canonical equations $v_\varphi = a\dot\varphi = \frac{a}{i\hbar}[\varphi, H]$ where the commutator takes the value $[\varphi, H] = i\hbar v_F a^{-1}\vec\sigma_\varphi \mathcal{I}_s$ and compute

$$J_Q = \frac{ev_F}{a} \sum_{m,\delta}' \left(\Psi_m^{\kappa,\delta}(\varphi, \Phi) \right)^\dagger \vec\sigma_\varphi \mathcal{I}_s \Psi_m^{\kappa,\delta}(\varphi, \Phi). \qquad (8.89)$$

Taking the ISO wave functions and substituting $m \to m + \Phi/\Phi_0$ we determine the appropriate $\Psi_m^{\kappa,\delta}(\varphi, \Phi)$. We could also, leave the wave function untouched and include a $U(1)$ gauge vector in the momentum operator. Let us explicitly write out an expectation value

$$J_{Q,\Delta}^{+,+} = -ev_F \left(\Psi_{m,\Delta}^{+,+}(\varphi, \Phi) \right)^\dagger \sigma_\varphi \mathcal{I}_s \Psi_{m,\Delta}^{+,+}(\varphi, \Phi) =$$

$$\frac{1}{4|E_{m,\Delta}^{+,+}|^2} \begin{pmatrix} \left[\epsilon(m - \frac{1}{2} + \frac{\Phi}{\Phi_0}) + i(\Delta_{SO} + E_{m,\Delta}^{+,+})\right]e^{i\varphi} \\ 0 \\ \epsilon(m - \frac{1}{2} + \frac{\Phi}{\Phi_0}) + i(\Delta_{SO} - E_{m,\Delta}^{+,+}) \\ 0 \end{pmatrix}^T$$

$$\begin{pmatrix} 0 & -ie^{-i\varphi} \\ ie^{i\varphi} & 0 \end{pmatrix} \otimes \begin{pmatrix} 1 & 0 \\ 0 & 1 \end{pmatrix}$$

$$(8.90)$$

$$
\begin{pmatrix}
\left[\epsilon(m - \tfrac{1}{2} + \tfrac{\Phi}{\Phi_0}) - i(\Delta_{SO} + E_{m,\Delta}^{+,+})\right] e^{-i\varphi} \\
0 \\
\epsilon(m - \tfrac{1}{2} + \tfrac{\Phi}{\Phi_0}) - i(\Delta_{SO} - E_{m,\Delta}^{+,+}) \\
0
\end{pmatrix}
$$

$$
= -\frac{\epsilon^2}{\Phi_0}\frac{(m - 1/2 + \Phi/\Phi_0)}{E_{m,\Delta}^{+,+}},
$$

which coincides with the expression of (8.85). With either of the two procedures one retrieves the same charge current of (8.85). This is a simple but interesting connection between linear response relations used to compute the current and a canonical exact calculation in principle. Note also that this expectation value corresponds to the procedure that eliminates Zitterbewegung from the Dirac definition of the velocity operator $\langle c\alpha \rangle$ where $\alpha = \sigma_\varphi \mathcal{I}_s$ and $c = v_F$. One can also obtain the linear response result using the group velocity operator applied to the free wave functions[80], where the group velocity operator is then

$$
\hat{J}_{Q,\Delta}^{\kappa,\delta} = \frac{v_F^2 \hat{\mathbf{p}}}{\kappa\sqrt{\Delta_{SO}^2 + \epsilon^2(m - \delta/2)^2}}, \tag{8.91}
$$

where $\hat{\mathbf{p}} = (-i\hbar/a)\partial_\varphi$. The first procedure above does not work for the Rashba coupling, that is, the mean value of the ordinary velocity operator in between the Rashba wave functions does not yield the linear response result. The second, group velocity approach depends on finding an appropriate Foldy-Wouthuysen transformation we believe is not currently known in the literature.

CHAPTER 9

CRYPTOGRAPHY AND QUANTUM MECHANICS

9.1 CLASSICAL CRYPTOGRAPHY

The first known use of cryptography can be found in an inscription on the main chamber of the tomb of Khnumhotep II, in Egypt, carved around 1900 BC. They used some unusual hieroglyphic symbols where the purpose was not to hide the message but perhaps to change its form in a way, which would make it appear dignified.

Evidence of some primary cryptography has been seen in most major early civilizations. "Arthshashtra," a classic work written in India by Kautalya, describes the espionage service and mentions assignments given to spies in "secret writing."

Ancients Romans were known to use a form of encryption to transmit secret messages to his army generals at the battle front. Known as Caesar cipher, is perhaps the most mentioned historic cipher in academic literature.

Is a substitution cipher, each character of a text is substituted by another character to form the cipher text. The variant used by Romans was a shift by 3 cipher. Each character was shifted by 3 places, so the character "A" was replaced by "D," "B" was replaced by "E," and so on.

Such ciphers depend only on the secrecy of the system, to keep safe the "key" and not on the encryption key itself. Once the system is known, the encrypted messages are easily decrypted. Substitution ciphers can be broken by using the frequency of letters in the language, brute force decryption.

At the beginning of the 19th century, when electricity was taking control over the industries and our way of living, Hebern, designed an electro-mechanical machine, which was called the Hebern rotor machine. The secret key was embedded in a rotating disc. The key encoded a substitution table and each key press from the keyboard resulted in the output of cipher text. This also rotated the disc by one notch and a different table would then be used for the next plain text character.

The Engima machine was invented by German engineer Arthur Scher-bius at the end of World War I, and was used by the German forces during the Second World War. The Enigma machine used 3 or 4 or even more rotors. The rotors rotate at different rates according to a keyboard typing and gave an output of appropriate cipher text letters. The key was the initial setting of the rotors.

Most of the work on cryptography was for military purposes, usually used to hide secret military information. However, cryptography attracted commercial attention in the post-war world, with businesses trying to secure their data from competitors.

Modernly, encryption or cryptographic systems are classified in three general dimensions:

First, the type of operations used for performing plaintext to cyphertext. All the encryption algorithms use of two general principles—substitution and transposition—through which plaintext elements are rearranged. It is relevant to say that no information should be lost.

Second, the number of keys used. If single key is used by both sender and receiver, it is called symmetric, single-key, secret-key or conventional encryption. If sender and receiver each use a different key, then it is called asymmetric, two-key or public-key encryption.

Third, the way in which plaintext is processed. A block cipher processes the input as blocks of elements and generates an output block for each input block. Cipher processes by stream, take the input elements continuously, producing output one element at a time as it goes along.

Finally, let us use one example in more detail, Caesar Ciphers. As we said above, Caesar cipher replaces each letter of the plaintext with an

a	b	c	d	e	f	g	h	i	j	k	l	m
0	1	2	3	4	5	6	7	8	9	10	11	12

n	o	p	q	r	s	t	u	v	w	x	y	z
13	14	15	16	17	18	19	20	21	22	23	24	25

Figure 9.1: Caesar cipher can be written as: $c = E(p) = (p + k)$ and $p = D(c) = (c - k)$. With a Caesar cipher, there are only 26 possible keys, of which only 25 are of any use.

alphabet. Examples can be given in the alphabet,

$$ABCDEFGHIJKLMNOPQRSTUVWXYZ.$$

Choose k, and shift all letters by k. For example, if $k = 5$, A becomes F, B becomes G, C becomes H, etc.

9.1.1 Encryption Security

Suppose a guy, Bob, wants to send his friend Alice a message via some insecure medium, and, to ensure that such messages are unreadable by anybody else, Bob and Alice have agreed previously that any messages they exchange will be encoded using the Caesar Cipher. Note that the medium itself doesn't matter at all: it could be a hand-penned message written on parchment and carried on horseback or it could be ASCII text sent as an e-mail.

The message that Bob wishes to send is called plaintext, the process of encoding it known as encryption and the resulting encoded message called ciphertext. To create ciphertext from plaintext using the Caesar Cipher is a simple task for Bob to perform. He merely replaces each letter in the message with the letter in the alphabet three places to its right - wrapping around back to 'A' after the letter 'Z'. So, a letter 'A' in the plaintext becomes a letter 'D' in the ciphertext, a 'B' becomes an 'E', and so on, all the way up to 'Z' which is replaced by a 'C'. Having encrypted his message, Bob then sends the ciphertext on its way to Alice, who, on receiving it, applies the reverse process and thus obtains the original plaintext message. To decrypt the encoded message, she replaces each letter in the ciphertext with the letter in the alphabet three places to its left.

Then, no matter how secure your encryption algorithm is, any conventional cryptographic methods suffer one major drawback. Since the same cryptographic key is used for encryption and decryption, for Alice to be able to read a message that Bob sends, Bob must give her the key that was used to encode the message. If this is performed over an insecure medium as well, then the key can easily be intercepted and the security of the message compromised without their knowledge.

This problem of "key management" was solved in 1976 with the invention of public key cryptography by Whitfield Diffie and Martin Hellman at Stanford University.

In public key cryptography, two mathematically related keys are generated for each party wishing to communicate, a public key and a private key. A message encoded using a particular public key can only be decrypted using the corresponding private key.

For Bob to send a message to Alice using public key cryptography, Bob encrypts the message using Alice's public key. Bob now transmit the ciphertext to Alice safely knowing that only Alice's private key can be used to decrypt and read it. Since her private key is used for decryption, Alice needs to keep this safe. But because she doesn't need to give this private key to anybody to establish a secure communication, there is no mayor problem in protecting her key. On the other hand, he needs to take no precautions with the public key. In fact, anybody can have access to it. He can post it, e-mail to friends or even publish it on the web.

Public key algorithms rely for their security on the fact that certain types of mathematical function are easy to perform in one direction but are difficult to do in reverse. For example, the popular RSA algorithm (named after its inventors Ronald Rivest, Adi Shamir, and Leonard Adlemann) makes use of the factorization problem: while it's easy to multiply two large prime numbers together, it is vastly more difficult to factor that product back into two large primes.

It is exactly this problem that would need to be solved for the codebreaker to be able to deduce an RSA private key form the corresponding public key. Other public key algorithms take advantage of other computationally hard problems, such as the discrete logarithm problem or elliptic curves.

The Internet is a completely public system, any information you send over the net is potentially visible to all.

The solution to security and privacy on the Internet is also public key cryptography.

Websites that employ the HTTPS protocol use the Secure Socket Layer (SSL) for authentication and encryption. It is public key cryptography, as the one used in software like PGP (Pretty Good Privacy), that safeguard the privacy of our e-mails and data, also, allows the creation of digital signatures, so that the authenticity and integrity of digital messages can be verified.

Most applications of cryptography on the Internet, such as SSL, work as an additional layer over existing Internet protocols. An increasingly popular application today is Virtual Private Networking (VPN), the creation of a secure, private network on top of an untrusted, public network like the Internet.

All data sent over a VPN is encrypted by IPSec, so the application software used to communicate over a VPN doesn't have to use or support cryptography for the data to remain secure.

As well as for encrypting data, public key cryptography can be used to create digital signatures. This signature can then be checked to ensure that the message or file was in fact created by the attributed author and that nobody has interfered with its contents since it was created. Digital signatures are widely used, for example, in the signing of software packaged for distribution on the Internet. They provide a level of confidence that the software you download hasn't been infected with a virus externally.

The first step in creating a digital signature is to use a cryptographic **hash** to generate a fingerprint of the message to be signed. This fingerprint is a sequence of bits, which uniquely or very nearly uniquely represents the message. A good hash function is designed to make it very difficult to generate another message, which would hash to the same sequence of bits, thus any modification of the message should result in a different fingerprint.

Following this, the fingerprint is encrypted with the author's private key to create the message signature. Both message and signature are then distributed.

Suppose Alice wishes to verify that a digitally signed message she has received was indeed sent by Bob. She uses Bob's public key to decrypt the message's digital signature and so reveal the message summary. She then computes a fresh message summary for the message that she received and compares it to the one that was encoded in the signature. If they don't match, either the message wasn't from Bob or the message has been altered with in some way.

One measure of an algorithm's security strength is the size of its key-space, the set of all possible keys that can be used with that algorithm.

If an algorithm's key-space is too small, then it is possible to break the encryption by brute force, we simply try each possible key and see if the ciphertext decrypts to a meaning text message.

More sophisticated algorithms can also be open to brute force attack especially with the available power of today's microprocessors.

The term **strong cryptography** is applied to algorithms that, even when the algorithm is known, are not sensitive to brute force attack.

Currently, block ciphers with a key size of 80 bits and above is considered secure for everyday purposes, while for public key systems such as RSA a safe key size is at least 1024 bits.

These key sizes are considered safe simply because there is currently insufficient computing power to exhaustively search such large key-spaces in a pragmatic time.

To crack strong cryptography, one must look for ways to reduce the number of keys that it is necessary search. This typically requires complex mathematical analysis, or cryptanalysis, of the particular algorithm to find its weakness, and several algorithms have been shown to be insecure in this way.

9.2 SCHRÖDINGER'S CAT

Erwin Schrödinger proposed an interesting model that illustrates the random nature of quantum mechanics. When we talk about Schrödinger's cat, we are referring to a paradox that arises from an imaginary experiment proposed by Erwin Schrödinger in 1935 [123] to illustrate the differences between interaction and measurement in quantum mechanics.

The mental experiment involves imagining a cat trapped inside a box that also contains a curious and dangerous device. This device is formed by a glass container, containing a very volatile poison and a hammer held over the container, so that if it falls on it breaks, and the poison escapes and the cat would die. The hammer is connected to an alpha particle detector mechanism. If an alpha particle arrives, the hammer falls, breaking the blister and the cat dies, on the contrary, if it does not arrive nothing happens and the cat is still alive.

When the whole device is ready, the experiment is performed. Next to the detector is a radioactive atom with certain characteristics: it has a 50% probability of emitting an alpha particle within an hour. Obviously, after one hour, one of the two possible events will have occurred: the atom has emitted an alpha particle or has not.

As a result of the interaction, inside the box, the cat is alive or dead. But we cannot know until we open it to prove it.

Figure 9.2: Schrödinger's cat and quantum superposition.

If we try to describe applying the laws of quantum mechanics what happens inside the box, we arrive to a peculiar conclusion. The cat will be described by an extremely complex wave function resulting from the superposition of two states combined at 50%: "alive cat" and "dead cat." That is, by applying quantum formalism, the cat would be both alive and dead. It would be in the two indistinguishable states.

The only way to find out what has happened to the cat is to open the box and look inside. In some cases we will find the cat is alive and others dead. When performing the measurement, the observer interacts with the system and "alters" it, breaks the overlapping states and the system "collapses" in one of its two possible states.

Common sense tells us that the cat cannot be alive and dead at the same time. But quantum mechanics says that as long as no one looks inside the box the cat is in a superposition of the two states: alive and dead.

This superposition of states is a consequence of the nature of matter and its application to a quantum mechanical description of physical systems, which allows to explain the behavior of elementary particles and atoms. The application to macroscopic systems like the cat would lead us to the paradox proposed by Schrödinger.

9.3   HEISENBERG UNCERTAINTY PRINCIPLE

We start by defining the commutator $[A, B]$ and anticommutator A, B between two operators A and B,

$$[A, B] \equiv AB - BA, \quad \{A, B\} \equiv AB + BA \tag{9.1}$$

If $[A, B] = 0$ then A commutes with B, if $A, B = 0$ then A anticommutes with B.

Suppose A and B are two hermitic operators and $|\psi\rangle$ is a quantum state.

Let's suppose $\langle \psi | AB | \psi \rangle = x + iy$ with $(x, y \in \mathbb{R})$. It is verified that $\langle \psi | [A, B] | \psi \rangle = 2iy$ and $\langle \psi | \{A, B\} | \psi \rangle = 2x$.

This implies that,

$$|\langle \psi | [A, B] | \psi \rangle|^2 + |\langle \psi | \{A, B\} | \psi \rangle|^2 = 4|\langle \psi | AB | \psi \rangle|^2 \tag{9.2}$$

By the Cauchy-Schwarz inequality,

$$|\langle \psi | AB | \psi \rangle|^2 \leq \langle \psi | A^2 | \psi \rangle \langle \psi | B^2 | \psi \rangle \tag{9.3}$$

Which combined with (9.2) and eliminating negative terms,

$$|\langle \psi | [A, B] | \psi \rangle|^2 \leq 4\langle \psi | A^2 | \psi \rangle \langle \psi | B^2 | \psi \rangle \tag{9.4}$$

Assuming that C and D are two observables. Substituting $A = C - \langle C \rangle$ and $B = D - \langle D \rangle$ in the last equation, we obtain the Heisenberg uncertainty principle as it is usually formulated,

$$\Delta(C)\Delta(D) \geq \frac{|\langle\psi|[C,D]|\psi\rangle|}{2} \tag{9.5}$$

In 1930 [124], Einstein showed that the uncertainty principle also implies the impossibility of reducing error in energy measurement without increasing the uncertainty of time during which the measure is performed. He believed this could be use as a springboard to refute the uncertainty principle, but Bohr proceeded to prove that Einstein's tentative refutation was erroneous.

In fact, the version of uncertainty, according to Einstein, proved to be very useful, because it meant that in a subatomic process the law on energy conservation could be violated briefly, provided that everything was restored to the state of conservation when these periods concluded. Yukawa used this notion to elaborate his theory of the pions. It made it possible to elucidate certain subatomic phenomena by assuming that particles arose out of nowhere as a challenge to conservation energy, but were extinguished before the time allotted to their detection, so they were only "virtual particles."

The "uncertainty principle" profoundly affected the thinking of physicists and philosophers. He had a direct influence on the philosophical question of "chance." But its implications for science are not what are commonly assumed.

Certainly, in many scientific observations, uncertainty is so insignificant compared to the corresponding scale of measures that it can be discarded for all practical purposes. One can simultaneously determine the position and motion of a star, or a planet, or a billiard ball, and even a grain of sand with absolutely satisfactory accuracy.

Regarding the uncertainty between the subatomic particles themselves, we can say that it is not an obstacle, but a real help for physicists. It has been used to elucidate facts about radioactivity, about the absorption of subatomic particles by nuclei, as well as many other subatomic events, with much more reasonableness than would have been possible without the uncertainty principle.

9.4   THE DENSITY OPERATOR

The language of the density operator provides a convenient means of describing quantum systems whose state is not fully known.   More precisely, suppose that a quantum system is in one state, $|\psi_i\rangle$, where i is a index that label the state, and where the respective probabilities are p_i.

We define as an assembly of pure states $\{p_i, |\psi_i\rangle\}$. Then is defined the density operator as,

$$\rho \equiv \sum_i p_i |\psi_i\rangle\langle\psi_i| \tag{9.6}$$

This operator is also known as the density matrix. All principles of quantum mechanics can be reformulated as a function of it. Suppose that the evolution of a closed quantum system is described by a unitary operator U. If the system is initially found in the state $|\psi\rangle$ with probability p_i, then after evolution the system will be in the state $U|\psi\rangle_i$ with equal probability p_i. Therefore, the evolution of the density operator can be described by,

$$\rho \equiv \sum_i p_i |\psi_i\rangle\langle\psi_i| \quad\xrightarrow{U}\quad \sum_i p_i\, U|\psi_i\rangle\langle\psi_i|U^\dagger = U\rho U^\dagger \tag{9.7}$$

Suppose that a measurement is performed by a "measurement operators" M_m. If the initial state was the ith, then the probability of obtaining the result m is the conditional probability,

$$p(m|i) = \langle\psi_i|M_m^\dagger M_m|\psi_i\rangle = \mathrm{Tr}\left[M_m^\dagger M_m|\psi_i\rangle\langle\psi_i|\right] \tag{9.8}$$

Considering the total probabilities, gives the absolute probability of obtaining the result m,

$$p(m) = \sum_i p(m|i)p_i = \mathrm{Tr}\left[M_m^\dagger M_m\rho\right] \tag{9.9}$$

And the state after the measurement is,

$$|\psi_i^m\rangle = \frac{M_m|\psi_i\rangle}{\sqrt{\langle\psi_i|M_m^\dagger M_m|\psi_i\rangle}} \tag{9.10}$$

Therefore after the measurement that generates the result m we have an assembly of states $|\psi\rangle_i^m$ with the respective probabilities $p(i|m)$. The density operator ρ_m is,

$$\rho_m = \sum_i p(i|m)|\psi_i^m\rangle\langle\psi_i^m| = \sum_i p(i|m)\frac{M_m|\psi_i\rangle\langle\psi_i|M_m^\dagger}{\langle\psi_i|M_m^\dagger M_m|\psi_i\rangle} \quad (9.11)$$

As it follows that $p(i|m) = p(m,i)/p(m) = p(m|i)pi/p(m)$ we obtain,

$$\rho_m = \sum_i p_i \frac{M_m|\psi_i\rangle\langle\psi_i|M_m^\dagger}{\text{Tr}\left[M_m^\dagger M_m \rho\right]} = \frac{M_m \rho M_m^\dagger}{\text{Tr}\left[M_m^\dagger M_m \rho\right]} \quad (9.12)$$

It is defined as a pure state the assembly formed by a single state, in this case $\rho = |\psi\rangle\langle\psi|$. Otherwise ρ is a mixed state, i.e., an assembly of pure states.

A simple criterion for knowing whether we have a pure or mixed state is to calculate its trace. A pure state satisfies $tr(\rho^2) = 1$, whereas for a mixed state $tr(\rho^2) < 1$ is verified.

Suppose that from an assembly $\{p_{ij}, |\psi_{ij}\rangle\}$ it is prepared a quantum system such that it is in the state ρ_i with probability p_i, then the density matrix results,

$$\rho = \sum_{ij} p_i p_{ij}|\psi_{ij}\rangle\langle\psi_{ij}| = \sum_i p_i \rho_i \quad (9.13)$$

An operator ρ is the density operator of an assembly $\{p_i, |\psi_i\rangle\}$ if these two conditions are satisfied,

- Unitary trace

$$\text{Tr}(\rho) = \sum_i p_i \text{Tr}\left[|\psi_i\rangle\langle\psi_i|\right] = \sum_i p_i = 1 \quad (9.14)$$

$$\langle\phi|\rho|\phi\rangle = \sum_i p_i|\langle\phi|\psi_i\rangle|^2 \geq 0 \quad (9.15)$$

- Positive definite

If the previous conditions are met, since ρ is positive, it has a spectral decomposition,

$$\rho = \sum_j \lambda_j |j\rangle\langle j| \tag{9.16}$$

where λ_j are the eigenvalues and $|j\rangle$ are the eigenvectors of ρ.

9.4.1 Reduced Density Operators and Partial Traces

To measure in a subsystems of a composite system, the only way to correctly describe observables is through the following definitions. Suppose that there are two physical systems A and B, whose states are described by an density operator ρ^{AB}. A reduced density operator for system A is defined as,

$$\rho^A = \text{Tr}_B\left[\rho^{AB}\right] \tag{9.17}$$

where tr_B is a map of operators known as the partial trace over the system B, and is defined as

$$\text{Tr}_B\left[|a_1\rangle\langle a_2| \otimes |b_1\rangle\langle b_2|\right] \equiv |a_1\rangle\langle a_2|\text{Tr}\left[|b_1\rangle\langle b_2|\right] = |a_1\rangle\langle a_2|\langle b_1|b_2\rangle \tag{9.18}$$

where $|a_1\rangle$, $|a_2\rangle$ are two arbitrary vectors of the state space of A and $|b_1\rangle$, $|b_2\rangle$ two arbitrary vectors of the state space of B. For instance, if a quantum system is in the product state $\rho^{AB} = \rho \otimes \sigma$ where ρ is the density operator of system A and σ corresponds to the system B, then,

$$\rho^A = \text{Tr}_B\left[\rho \otimes \sigma\right] = \rho\text{Tr}(\sigma) = \rho \tag{9.19}$$

which is intuitive.

In the same way $\rho^B = \sigma$. We deduce from this that the reduced density operator of a composite system is a pure state. A second example is the Bell's state $(|00\rangle + |11\rangle)/\sqrt{2}$. In this case, the density operator will be,

$$\rho = \left(\frac{|00\rangle + |11\rangle}{\sqrt{2}}\right)\left(\frac{\langle 00| + \langle 11|}{\sqrt{2}}\right) \tag{9.20}$$

$$= \frac{1}{2}\left(|00\rangle\langle 00| + |11\rangle\langle 00| + |00\rangle\langle 11| + |11\rangle\langle 11|\right) \tag{9.21}$$

Looking for the partial density operator for the first,

$$\rho^1 = \text{Tr}_2(\rho)$$

$$= \frac{1}{2} \left(\text{Tr}_2[|00\rangle\langle00|] + \text{Tr}_2[|11\rangle\langle00|] + \text{Tr}_2[|00\rangle\langle11|] + \text{Tr}_2[|11\rangle\langle11|] \right)$$

$$= \frac{1}{2} \left(|0\rangle\langle0|\langle0|0\rangle + |1\rangle\langle0|\langle0|1\rangle + |0\rangle\langle1|\langle1|0\rangle + |1\rangle\langle1|\langle1|1\rangle \right)$$

$$= \frac{1}{2} \left(|0\rangle\langle0| + |1\rangle\langle1| \right)$$

$$= \frac{1}{2} \begin{pmatrix} 1 & 0 \\ 0 & 1 \end{pmatrix} \tag{9.22}$$

We note that the Bell pair is a mixed state, since $tr((\sqrt{2}/2)^2) = \frac{1}{2} < 1$.

9.4.2 Schmidt's Decomposition and Purifications

The reduced density operators and partial traces are the basis of a wide variety of useful tools for the study of compound quantum systems. Two of those tools are Schmidt's decomposition and purifications.

9.4.2.1 Schmidt's Decomposition

Suppose that $|\psi\rangle$ is the pure state of a composite system AB. Then there are orthonormal states $|i_A\rangle$ for an A system and orthonormal states $|i_B\rangle$ for a system B, such that

$$|\psi\rangle = \sum_i \lambda_i |i_A\rangle \otimes |i_B\rangle \tag{9.23}$$

where λ_i are non-negative real numbers, known as Schmidt coefficients. These are such that satisfy $\sum_i \lambda_i^2 = 1$.

For example, for the previous system it follows that,

$$\rho^A = \sum_i \lambda_i^2 |i_A\rangle\langle i_A|, \qquad \rho^B = \sum_i \lambda_i^2 |i_B\rangle\langle i_B| \tag{9.24}$$

That is, the eigenvalues of both reduced density operators are equal. If applied to a simple system like,

$$|\psi\rangle = \frac{1}{\sqrt{3}} \left(|00\rangle + |01\rangle + |11\rangle \right) \tag{9.25}$$

which is not symmetrical, but calculating the reduced traces, we obtain coincident results,

$$\text{Tr}\left[\left(\rho^A\right)^2\right] = \text{Tr}\left[\left(\rho^B\right)^2\right] = \frac{7}{9} \tag{9.26}$$

9.4.2.2　*Purification*

Supposing that ρ^A is the state of a quantum system A. It is possible to introduce another system that we will call R, and define a pure state for the composite system $|AR\rangle$ such that $\rho^A = Tr_R(|AR\rangle\langle AR|)$. This means that the pure state $|AR\rangle$ is reduced to ρ^A when we focus only on the system A. R is a fictitious system, and is known as a reference system.

Let us suppose that ρ^A has normal orthonormal decomposition $\rho^A = \sum_i p_i |i^A\rangle\langle i^A|$. To purify ρ^A we introduce a reference system R that has the same space of states of A, with orthonormal bases $|i^R\rangle$ and define a pure state for the combined system,

$$|AR\rangle \equiv \sum_i \sqrt{p_i}|i^A\rangle \otimes |i^R\rangle \tag{9.27}$$

Calculating the reduced density operator. We verify that $|AR\rangle$ is a purification of ρ^A,

$$\begin{aligned}
\text{Tr}_R\left[|AR\rangle\langle AR|\right] &= \sum_{ij}\sqrt{p_i p_j}|i^A\rangle\langle j^A|\text{Tr}\left[|i^R\rangle\langle j^R|\right] \\
&= \sum_i p_i |i^A\rangle\langle i^A| = \rho^A
\end{aligned} \tag{9.28}$$

Attention is payed to the close relationship between Schmidt's decomposition and the purification process: the method used to purify a mixed state of a system A is to define a pure state whose Schmidt base for A is just the basis on which the mixed state is diagonal, and in which the Schmidt coefficients are the square roots of the eigenvalues of the purified density operator.

9.4.3　EPR, Bell's Inequality, and the Violation of Local Realism

The essence of the EPR argument is as follows. Einstein, Podolsky and Rosen were interested in what they called "elements of reality." The had the conviction that any of these elements must be present in any complete physical theory. The EPR goal was to show that quantum mechanics lacked of some elements of reality and therefore was incomplete.

The path pursued was the introduction of a "sufficient property" of an observable to be cataloged as an element of reality, and that property was the ability to predict with certainty the value that the observable will have in the future, but a moment before being measured.

Considering a couple of what we will call from now, quantum bits of information (qubits), anticorrelated in the form,

$$\frac{1}{\sqrt{2}} \left(|01\rangle - |10\rangle \right) \tag{9.29}$$

The two components are separated and the qubits are placed in the hands of two observers A and B, that are widely separated in space and time. Suppose A performs on its qubit a measure of spin along the axis $\hat{v}$, that is to say A measures the observable $\hat{v} \cdot \vec{\sigma}$ and the result obtained is $+1$. Thus, a simple quantum mechanical calculation allows to predict with certainty that B will measure -1, if an equivalent measurement is performed. Since A can predict the result of the measure of B, the spin measurement along the $\hat{v}$-axis becomes an element of reality according to the EPR criterion. However, quantum mechanics as presented is limited to calculate the probabilities of the respective results when $\hat{v} \cdot \vec{\sigma}$ is measured.

Thirty years after the EPR publication, an experimental test was performed to verify the existence of the intuition advocated by EPR against quantum mechanics. Experiments invalidated the EPR suggestion and provided support for quantum mechanics. The key to this experimental invalidation is known as Bell's inequality. Let's first look at the classical view of the experiment and then the similar quantum experiment.

Classical View   An independent observer C prepares two particles and distributes each to a couple of observers A and B. A measures two physical properties P_Q and P_R, respectively, whose results can be $+1$ and -1. The A particle has the value Q for the property P_Q and it is assumed that Q is an objective and real property that is simply revealed by the measurement. Similarly R is the value of the other property P_R and the result of its measurement. Similarly, B is able to measure one of two properties P_S and P_T, which would reveal the real values S and T. Both choose at randomly a single property each. Now we assume that A and B are distanced light years and "simultaneously" perform their measurements so that the result obtained by A cannot affect the result of the measurement of B, since the physical influences cannot propagate at a faster rate than light in the Vacuum.

The four cases of mutually exclusive measurements. If we generate the sum function $QS + RS + RT - QT = (Q + R)S + (R - Q)T$ and given that R, Q are valued ± 1, or $(Q + R)$ and/or $(R - Q)$ are zero, which result that the sum function will be ± 2.

Let $p(q, r, s, t)$ be the a priori probability that the system is in the state $Q = q$, $R = r$, $S = s$, $T = t$. These probabilities will depend on how C has prepared the particles and the experimental fluctuations. Calculating the expectation value of the sum function, the results are,

$$E(QS + RS + RT - QT) =$$
$$= \sum_{qrst} P(q, r, s, t)(qs + rs + rt - qt) \leq \sum_{qrst} P(q, r, s, t) = 2$$
$$E(QS) + E(RS) + E(RT) - E(QT) \leq 2 \tag{9.30}$$

This last equation is known as the Bell inequality. By repeating the experiment several times, A and B can determine each value on the left side hand of (9.30).

Quantum Vision Investigator C prepares a quantum system of two correlated or entangled states,

$$|\psi\rangle = \frac{1}{\sqrt{2}} (|01\rangle - |10\rangle) \tag{9.31}$$

and passes each qubits to A and B.

They perform measurements of the following observables,

$$Q = Z_1, \quad S = -\frac{1}{\sqrt{2}}(Z_2 + X_2),$$
$$R = X_1, \quad T = \frac{1}{\sqrt{2}}(Z_2 - X_2) \tag{9.32}$$

The averages are

$$\langle QS \rangle = \sqrt{2}, \quad \langle RS \rangle = \sqrt{2}, \quad \langle RT \rangle = \sqrt{2}, \quad \langle QT \rangle = -\sqrt{2} \quad (9.33)$$

then

$$\langle QS \rangle + \langle RS \rangle + \langle RT \rangle - \langle QT \rangle = 2\sqrt{2} \quad (9.34)$$

The comparison between (9.30) and (9.33) shows a contradiction that must be solved experimentally. Optical experiments confirmed quantum prediction in detriment of classical. Nature does not obey Bell's inequality. This means that one or more premises used in the deduction of (9.30) must be incorrect. There are two implicit bases in (9.30) that may seem doubtful, the premise that indicates the properties P_Q, P_R, P_S, P_T have defined values Q, R, S, T regardless of their observation; and a premise that indicates the measurement of A does not affect the measurement of B, this is called locality. Both combined assumptions are known as the Local Realism model. These are absolutely intuitive and reasonable assumptions about how our universe works and fit well into our everyday experience. However, the failure of Bell's inequality shows us that at least one of them is incorrect.

9.5 QUANTUM COMPUTATION

The spectacular promise of quantum computers consists in the development of a set of new algorithms that make feasible problems that require exorbitant resources for their resolution in traditional computers. Two kinds of quantum algorithms are known that fulfill this promise. The first is based on Shor's quantum Fourier transform, encompassing remarkable algorithms for factoring and finding discrete logarithms that are exponentially faster than the best known ones. The second class is based on Grover's quantum quest algorithm. These provide a remarkable acceleration of quadratic order on the best traditional algorithms. The latter method is particularly important because of the wide use of in search methods. The following figure shows the current state of the families of known quantum algorithms

The problem of the discrete logarithm together with the problem of determining the order of a multiplicative group such as that presented in modular exponentiation. Both techniques allow solving the problem of factorization, the basis of many public key cryptosystems such as the RSA (Rivest-Shamir-Adleman) algorithm for security in the SSL protocol of TCP/IP (Internet) networks, virtual private networks (VPN), digital signature and digital e-commerce certificates and many other fields of enormous interest in the field of computer security.

The problem of the discrete logarithm allows breaking cryptosystems, linked to algebraic multiplicative groups such as elliptic curves and Galois fields. In turn, the quantum search algorithm solves statistical problems in vast databases and break symmetric cryptosystems by exhaustively exploring the key space.

Finding a good quantum algorithm seems to be a complex problem. There are at least two reasons for this, on the one hand the design of quasi-optimal algorithms is complex even for seemingly simple problems, like multiplying two numbers, and finding good quantum algorithms imposes the additional constraint of having to overcome classical algorithms. The other reason is that our intuition is better adapted to the classical world than to the quantum world. Then a critical analysis of the components of a quantum algorithm will be made to know its possibilities and limitations.

9.5.1 Operations on Isolated Qubits

The development of quantum computational tools starts with operations on the simplest elements, the qubits. Operations are described in terms of logical gates. These ports fulfill the equivalent function of Boolean operators such as NOT gates. In classical computing The combination of these logic gates allow the development of quantum algorithms to be transferred to hardware if the conditions exist.

A qubit is a vector $|\psi\rangle = a|0\rangle + b|1\rangle$ the orthonormal base $|0\rangle, |1\rangle$ and characterized by the complex parameters a, b satisfying $|a|^2 + |b|^2 = 1$. In addition each qubit can be represented in the Bloch sphere through the change of coordinates $a = \cos\theta/2$, $b = e^{i\phi}\sin\theta/2$. Logical gates must preserve this rule and therefore must be $2 times 2$ unit operators. We will name unitary to 2×2 unit operators that have a single input and output qubit. The most important endowment of the logic gates is the following set of six operators, the Pauli Gates (matrices),

$$X = \begin{pmatrix} 0 & 1 \\ 1 & 0 \end{pmatrix}, \quad Y = \begin{pmatrix} 0 & -i \\ i & 0 \end{pmatrix}, \quad X = \begin{pmatrix} 1 & 0 \\ 0 & 1 \end{pmatrix} \tag{9.35}$$

$$H = \frac{1}{\sqrt{2}} \begin{pmatrix} 1 & 1 \\ 1 & -1 \end{pmatrix}, \quad S = \begin{pmatrix} 1 & 0 \\ 0 & i \end{pmatrix}, \quad X = \begin{pmatrix} 1 & 0 \\ 0 & e^{i\frac{\pi}{4}} \end{pmatrix} \tag{9.36}$$

the following relations between them are verified,

$$H = \frac{1}{\sqrt{2}}(X + Z), \quad S = T^2 \tag{9.37}$$

Pauli matrices give rise to three classes of unit matrices, when they are exponentiated,

$$R_x(\theta) = e^{-i\frac{\theta}{2}X} = \cos\left(\frac{\theta}{2}\right)\mathbb{I} - i\sin\left(\frac{\theta}{2}\right)X = \begin{pmatrix} \cos\left(\frac{\theta}{2}\right) & -i\sin\left(\frac{\theta}{2}\right) \\ -i\sin\left(\frac{\theta}{2}\right) & \cos\left(\frac{\theta}{2}\right) \end{pmatrix}$$

$$R_y(\theta) = e^{-i\frac{\theta}{2}Y} = \cos\left(\frac{\theta}{2}\right)\mathbb{I} - i\sin\left(\frac{\theta}{2}\right)Y = \begin{pmatrix} \cos\left(\frac{\theta}{2}\right) & -\sin\left(\frac{\theta}{2}\right) \\ \sin\left(\frac{\theta}{2}\right) & \cos\left(\frac{\theta}{2}\right) \end{pmatrix}$$

$$R_z(\theta) = e^{-i\frac{\theta}{2}Z} = \cos\left(\frac{\theta}{2}\right)\mathbb{I} - i\sin\left(\frac{\theta}{2}\right)Z = \begin{pmatrix} e^{-i\frac{\theta}{2}} & 0 \\ 0 & e^{i\frac{\theta}{2}} \end{pmatrix} \tag{9.38}$$

Generating the rotation operators around the axes $\hat{x}$, $\hat{y}$ and $\hat{z}$. The interpretation of these operators on the Bloch' sphere comes as if we have qubit in three dimensions such that $\hat{n} = (n_x, n_y, n_z)$ and one of the rotation operators is applied with an angle θ, the vector rotated around the corresponding axis θ degrees. To simplify circuits it is convenient to remember that,

$$HXH = Z, \quad HYH = -Y, \quad HZH = X \tag{9.39}$$

The basic properties of quantum circuits are as follows: time evolves from left to right, single lines or connections represent qubits and a "/" represents a set of qubits. The basic layout of gates is,

9.5.2　Operations on Multiple Qubits

One of the most useful operations in both classical and quantum computing is the conditional bifurcation of type "If A is true then run B." These are

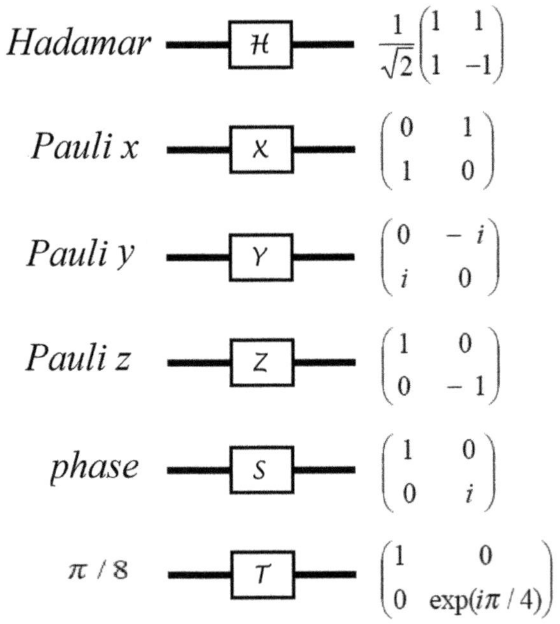

$$Hadamar \quad -\boxed{\mathcal{H}}- \quad \frac{1}{\sqrt{2}}\begin{pmatrix} 1 & 1 \\ 1 & -1 \end{pmatrix}$$

$$Pauli\ x \quad -\boxed{X}- \quad \begin{pmatrix} 0 & 1 \\ 1 & 0 \end{pmatrix}$$

$$Pauli\ y \quad -\boxed{Y}- \quad \begin{pmatrix} 0 & -i \\ i & 0 \end{pmatrix}$$

$$Pauli\ z \quad -\boxed{Z}- \quad \begin{pmatrix} 1 & 0 \\ 0 & -1 \end{pmatrix}$$

$$phase \quad -\boxed{S}- \quad \begin{pmatrix} 1 & 0 \\ 0 & i \end{pmatrix}$$

$$\pi/8 \quad -\boxed{T}- \quad \begin{pmatrix} 1 & 0 \\ 0 & \exp(i\pi/4) \end{pmatrix}$$

Figure 9.3: Logical gates for isolated qubits.

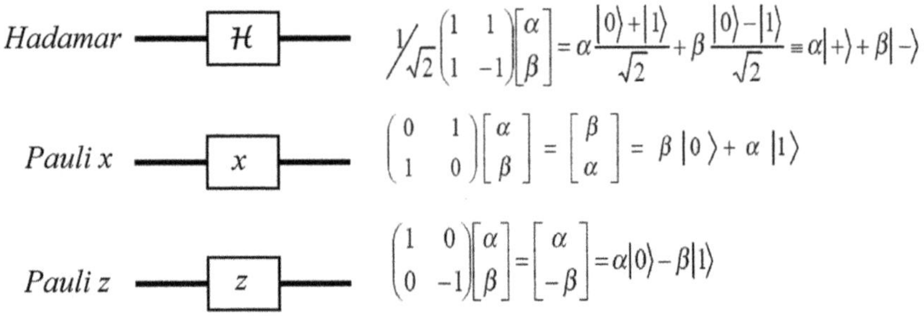

$$Hadamar \quad -\boxed{\mathcal{H}}- \quad \frac{1}{\sqrt{2}}\begin{pmatrix} 1 & 1 \\ 1 & -1 \end{pmatrix}\begin{bmatrix} \alpha \\ \beta \end{bmatrix} = \alpha\frac{|0\rangle+|1\rangle}{\sqrt{2}} + \beta\frac{|0\rangle-|1\rangle}{\sqrt{2}} \equiv \alpha|+\rangle + \beta|-\rangle$$

$$Pauli\ x \quad -\boxed{x}- \quad \begin{pmatrix} 0 & 1 \\ 1 & 0 \end{pmatrix}\begin{bmatrix} \alpha \\ \beta \end{bmatrix} = \begin{bmatrix} \beta \\ \alpha \end{bmatrix} = \beta|0\rangle + \alpha|1\rangle$$

$$Pauli\ z \quad -\boxed{z}- \quad \begin{pmatrix} 1 & 0 \\ 0 & -1 \end{pmatrix}\begin{bmatrix} \alpha \\ \beta \end{bmatrix} = \begin{bmatrix} \alpha \\ -\beta \end{bmatrix} = \alpha|0\rangle - \beta|1\rangle$$

Figure 9.4: Three of the most used isolated qubit operations.

binary operations in the sense that they operate on two qubits and therefore are 4×4 matrices. The prototype of this class of operators is the CNOT, the top line represents the control qubit, the lower one to the controlled qubit. The operation is as follows, if t is the target qubit and c is the control qubit, then $|c\rangle|t\rangle \rightarrow |c\rangle|c \oplus t\rangle$, if c is $|1\rangle$ flips its state to $|0\rangle$, otherwise t is not modified. Here we show the CNOT diagram,

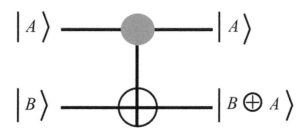

Figure 9.5: CNOT gate—the top line is the control qubit, and the lower is the qubit that target next to its matrix.

A very important feature that distinguishes the classic CNOT gate from quantum is that the former allows "duplicate" bits, whereas quantum states cannot be duplicated or cloned.

In general, if U is an arbitrary operator over unit qubits, the U-CONTROLLED (UC) gate can be created in a similar way to the previous one, it only acts U if c is $|1\rangle$,

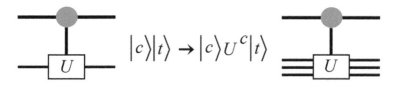

Figure 9.6: U-Controlled gate—the upper qubit is the control operator and to the right is the multiple control gate.

with

$$F(a) = \begin{pmatrix} 1 & 0 \\ 0 & e^{ia} \end{pmatrix} \qquad (9.40)$$

An object of interest is to find an equivalent circuit to U-CONTROLLED (UC), based exclusively on unary operators. To do this, we resort to the following equivalence. Let be U an unit operator over

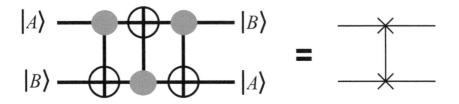

Figure 9.7: Swap gate.

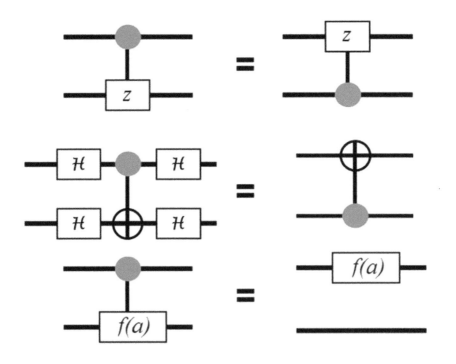

Figure 9.8: Gate equivalences.

isolated qubits, then it can be shown that there are unit operators A, B, 4 on isolated qubits, such that $ABC = \mathcal{I}$, $U = exp(i\alpha)A \times B \times C$ and α is general phase factor, this results

In general the circuit may have more than two control lines, if it has n control qubits operating on a unary gate U of one qubit, it would be defined as a gate $C^n(U)$.

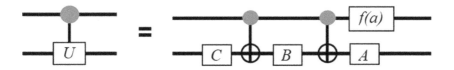

Figure 9.9: Equivalence of the U-CONTROLLED (UC) gate as a function of unary operators.

Finally we have the operator or measuring gate, which transforms a qubit into a classic bit. The following diagram presents a measuring gate, the double lines are used to represent classic bits

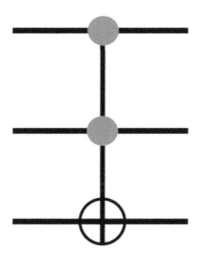

Figure 9.10: Toffoli gate.

A Toffoli gate can be constructed from 4 unitary operators

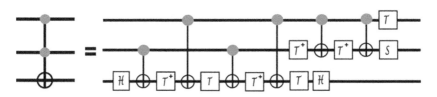

Figure 9.11: Sequential decomposition, Toffoli gate.

The measure operator transforms a qubit into a classical bit.

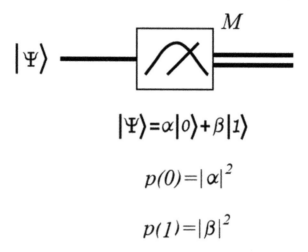

$$|\Psi\rangle = \alpha|0\rangle + \beta|1\rangle$$

$$p(0) = |\alpha|^2$$

$$p(1) = |\beta|^2$$

Figure 9.12: Measure operator.

9.5.3   Quantum Parallelism

Quantum parallelism is one of the fundamental attributes of quantum algorithms and the main responsible for its enormous computational potential, compared to its classical counterpart.

Heuristically, quantum parallelism allows quantum computers to evaluate a function $f(x)$ for multiple values of x simultaneously.

Let $f(x) : 1,0 \rightarrow 1,0$ be a function with a bit as domain and rank. A U_f transformation is defined by the map $|x, y\rangle \rightarrow |x, y \oplus f(x)\rangle$, where $\oplus$ is the sum module two. Assuming that for us the detailed circuit of is a black box, we symbolize it as,

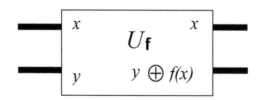

Figure 9.13: U_f operator of parallel evaluation of $f(0)$ and $f(1)$.

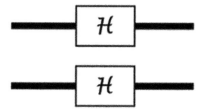

Figure 9.14: Hadamard transform $H^{\otimes 2}$ operating on two qubits in parallel.

We can create a superposition basis $\frac{|1\rangle + |2\rangle}{\sqrt{2}}$, with the Hadamard transformation operating on a $|0\rangle$. If that base is entered in the circuit described above next to a null qubit ring, it results

$$U_f\left(\left|\frac{|0\rangle + |1\rangle}{\sqrt{2}}, |0\rangle\right\rangle\right) \quad \rightarrow \quad \left|\frac{|0, f(0)\rangle + |1, f(1)\rangle}{\sqrt{2}}\right\rangle \tag{9.41}$$

The resulting superposed state contains in a single qubit the evaluations of both $f(0)$ and $f(1)$. This is the essence of quantum parallelism. Contrary to classical circuits where parallelism is achieved by multiple circuits each computing $f(x)$ for different x, here a single circuit and a single evaluation computes both values.

This procedure is generalized for an arbitrary number of qubits by the Hadamard or Walsh-Hadamard Transform, which simply consists of the application of n gates H in parallel.

For example, in this case, if the inputs are zero states,

$$|00\rangle \xrightarrow{H^{\otimes 2}} \left(\frac{|0\rangle + |1\rangle}{\sqrt{2}}\right) \otimes \left(\frac{|0\rangle + |1\rangle}{\sqrt{2}}\right) = \frac{1}{\sqrt{2}}\left(|00\rangle + |01\rangle + |10\rangle + |11\rangle\right) \tag{9.42}$$

In general, if we enter n qubits in null state, the output state will be

$$H^{\otimes n} = \frac{1}{\sqrt{2^n}} \sum_{x} |x\rangle \tag{9.43}$$

Where the sum covers all possible quantum states of the set of qubits. The Hadamard transform generates the symmetric overlap of all base states using only n logic gates, which is very efficient. For example, to evaluate a function $f(x)$ as defined above, parallel over all its domain values, first the state is prepared $|0\rangle^{\oplus n} = |000...0\rangle_n$, it is processed through the Hadamard transform to superpose all states, and finally, next to a qubit null ring is passed through a U_f transform, the result will be

$$U_f \left(|00 \cdots 0\rangle_n \right) \quad \rightarrow \quad \frac{1}{\sqrt{2^n}} \sum_x |x\rangle |f(x)\rangle \qquad (9.44)$$

Where as before the sum covers all the superposed states of the n qubits. In a sense, the quantum parallelism has allowed to evaluate simultaneously the function $f(x)$ on the 2^n combinations of superposed states of n qubits.

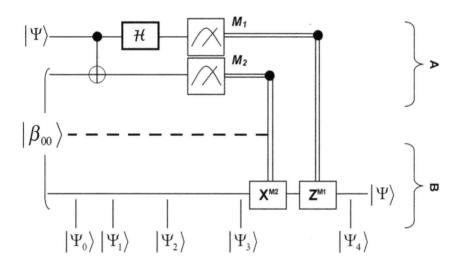

Figure 9.15: Teleportation circuit of an unknown quantum qubit state from Alice to Bob by transmission of two classic bits. Both have a half-grid input. In the lower row the successive quantum states of the process.

9.5.4   Quantum Teleportation

Suppose Alice and Bob have known each other for some time and generate entangled qubit $|\beta_{00}\rangle$. Then, they are separated and each one gets an entangled qubit. In a given instant, Alice decides to send Bob a qubit $|\psi\rangle$, whose state is unknown to her. She only has the possibility of send Bob classic bits, and as they both have quantum hardware, they run the following experiment, see Figure 9.14.

We can follow the successive states by applying step by step the following operators,

State to teleport:

$$|\psi\rangle = \alpha|0\rangle + \beta|1\rangle \tag{9.45}$$

Starting state:

$$|\psi_0\rangle = |\psi\rangle \otimes |\beta_{00}\rangle = \frac{1}{\sqrt{2}}\left[\alpha|0\rangle(|00\rangle + |11\rangle) + \beta|1\rangle(|00\rangle + |11\rangle)\right] \tag{9.46}$$

CNOT:

$$|\psi_1\rangle = \frac{1}{\sqrt{2}}\left[\alpha|0\rangle(|00\rangle + |11\rangle) + \beta|1\rangle(|10\rangle + |01\rangle)\right] \tag{9.47}$$

HADAMARD:

$$\begin{aligned}
|\psi_2\rangle &= \frac{1}{2}\left[\alpha(|0\rangle + |1\rangle)(|00\rangle + |11\rangle) + \beta(|0\rangle - |1\rangle)(|10\rangle + |01\rangle)\right] \\
&= \frac{1}{2}\left[|00\rangle(\alpha|0\rangle + \beta|1\rangle) + |01\rangle(\alpha|1\rangle + \beta|0\rangle) \right. \\
&\quad \left. + |10\rangle(\alpha|0\rangle - \beta|1\rangle) + |11\rangle(\alpha|1\rangle - \beta|0\rangle)\right]
\end{aligned} \tag{9.48}$$

As we can see, when leaving Hadamard's gate, the state separates into four independent and orthogonal terms. For example, the first term has the qubits of Alice in state $|00\rangle$ and when measures obtains 00, Bob's qubit will be in the state $\alpha|1\rangle + \beta|0\rangle$, i.e., the state of the teleported qubit. Logically, Alice's qubits can be in one of four orthogonal states and therefore there are four possible outcomes of the experiment,

$$\begin{aligned}
\text{if} \quad a = b = 0 &\Longrightarrow |\psi_4\rangle = \alpha|0\rangle + \beta|1\rangle \\
\text{if} \quad a = b = 1 &\Longrightarrow |\psi_4\rangle = \alpha|1\rangle - \beta|0\rangle \\
\text{if} \quad a = 0\,, b = 1 &\Longrightarrow |\psi_4\rangle = \alpha|1\rangle + \beta|0\rangle \\
\text{if} \quad a = 1\,, b = 0 &\Longrightarrow |\psi_4\rangle = \alpha|0\rangle - \beta|1\rangle
\end{aligned} \tag{9.49}$$

9.5.5 Superdense Encoding

In a sense it can be considered an inverse experiment to teleportation, which is emphasized when analyzing the entropic balance of both phenomena. In this case, Alice and Bob share a Bell qubit entangled $|\beta_{00}\rangle$. Alice wants to send Bob in a single qubit two classic bits of information, hence the name of *superdense informational encoding*. This works in the following way, Alice gives back to Bob her entangled qubit, but choose one of four options, based on Bell's defined operators, before submitting it, for example, if you choose option 01 transform the qubit by the Z gate before returning it, in general,

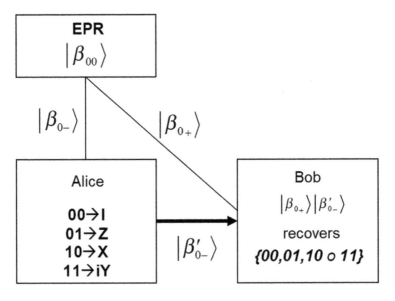

Figure 9.16: Scheme of the superdense coding (Alice decides to communicate to Bob 2 classic bits and does so by sending only 1 qubit).

9.5.5.1 *Feynman Diagrams and Entropic Balance of Teleportation and Superdense Coding*

In the late 1940s, Feynman stated that quantum-mechanical results could be represented as the sum of all possible paths joining an initial point in space and time, with an end point also in space and time [125]. Each of these paths contributes in a special way and carries a characteristic phase. In order to find the result it is necessary to take into account each one of the phases, because if one path would sum in destructive or constructive interference.

This formulation is used in high-energy physics. The difficulty always lies in the need to find a very large number, in fact infinite, of paths, and to

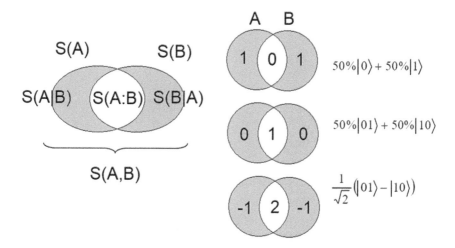

Figure 9.17: Entropy balance of a two-qubit system (On the left the Venn diagram of entropy in a bipartite AB system, and on the right the diagram of two qubits A and B with S (A) = 1 and S (B) = 1 in three limit states).

calculate the phases in detail in order to be able to add them up and obtain the desired result.

In our case, the diagrams of Feynman allow to represent in the simplest possible way teleportation phenomena and superdense coding. At the same time, it is extremely interesting to analyze from the entropic point of view both phenomena, in particular the existence of negative conditional entropies.

The entropies of Von Neumann of the different states are,

$$S(A) = -\text{Tr}\left[\rho_A \log_2 \rho_A\right] , \quad S(B) = -\text{Tr}\left[\rho_B \log_2 \rho_B\right] \tag{9.50}$$

where

$$S(A|B) = -\text{Tr}\left[\rho_{AB} \log_2 \rho_{AB}\right] \tag{9.51}$$

$$\rho_{AB} = \lim_{n \to \infty} \left[\rho_{AB}^{1/n} \left(\mathbb{I} \otimes \rho_B\right)^{-1/n}\right]^{1/n} \tag{9.52}$$

$$S(A:B) = -\text{Tr}\left[\rho_{AB} \log_2 \rho_{A:B}\right] \tag{9.53}$$

$$\rho_{A:B} = \lim_{n \to \infty} \left[(\rho_A \otimes \rho_B)^{1/n} \rho_{AB}^{-1/n} \right]^n \tag{9.54}$$

ρ_{AB} occupies a special role because it is not semi-define positive, so $S(A|B)$ can be negative, a fact that is fulfilled in the case of entangled states. We define as "virtual information" to that in which its extraction or recovery would violate the principle of causality.

In other words, virtual information cannot result in superluminal communications between the two particles of the EPR pair. The sign of the virtual information that is assigned to each element of the EPR pair is not an intrinsic property of the particle, but it is only assignable once the dynamics process has been completed.

CHAPTER 10

NON-EQUILIBRIUM QUANTUM MECHANICS

10.1 SYSTEMS IN EQUILIBRIUM

The assumption that a system is in equilibrium is very powerful. It allows a massive reduction of the mathematical complexity required to describe a system. It is also a very restrictive statement with needs that are never completely satisfied. We outline exactly where these assumptions play a role and how they are used to make more manageable problems apparently too complex.

10.1.1 Open and Closed Systems

There are many definitions of what is balance. It is define as the simultaneous satisfaction of two assumptions. The first is that a balanced system is characterized by a unique set of variables, extensive and intensive, that does not change over time. The second is that after isolation of the system, all variables remain unchanged. This assumption has already distributed the total system of in a "system" and "environment" [115], and most of work is directed to assuming that the system is characterized by extensive and intensive variables independent of time. We immediately know that this definition is incomplete because these thermodynamic variables are what we need for defining the equilibrium. The thermodynamic variables could be defined as stationary, but when the potential is switched off, the distribution of particles will relax into a Gibbs distribution.

The thermodynamic variables will also change. The second hypothesis defines these cases as non-equilibrium. This situation also could be considered a balance if the potential is never turned off. In the same way, an ideal gas contained in a volume can be considered non-equilibrium if any change is made in the volume in the distant future. The point of these considerations is to realize that whether the system is or not in equilibrium depends critically on what is called the "system" and its surrounding "environment."

Let's discuss how a total system partition occurs. Consider what can be exchanged among lots of elementary excitations or particles. For example, a gas in a container. In this very abstract example, take into account that the container is made of an infinitely rigid material that does not change due to collisions with the gas particles. This gas can be considered an isolated system, a system which does not exchange energy and matter with any external environment [1].

Consider a gas of N atoms of a fissile nuclear material. We perform a statistical analysis of the average values of observables such as position and momentum. If we limit ourselves to time scales that are short compared with the half-life of atoms, there is no problem in defining an isolated system. The number of particles N will remain constant, and the system can be modeled as if it were isolated. However, if we consider the system at times that are long compared with the half-life, then $N \to 0$ as the decay occurs. The system should be modeled as fully open with a "system" exchanging atoms of matter and energy with an "environment."

Fundamentally, it is how the system is observed and described that determines whether it is or not opened or isolated.

There are other ways in which a system can interact with an environment. It can exchange energy without exchanging matter. This system is called "closed." A thermos of coffee can be modeled as isolated if we only consider times that are short compared to the time it takes to cool down, thereby releasing its heat into the surrounding air. The dissipation of energy in a thermal reservoir that dissipates energy are sufficiently general to describe a series of interesting phenomena.

There are other ways to model the interactions of systems with an environment. One could envision a system designed so that matter but no energy is exchanged, by requiring that the particles are removed or added only when their energies are zero.

10.1.2 Equilibrium

We will begin by analyzing a system K. The system consists in a set of microscopic degrees of freedom, which can be found in certain quantum States. If we know the energy of every degree of freedom, we may say that we know the microstate of the system. For example, if K represents an isolated gas and we knew the quantum state of every particle of gas, then the set α_i, where α_i is a collective index of quantum numbers, represents a microstate of K. The system will always be instantly in a microstate, but due to the large number of microscopic degrees of freedom it would be impossible to write the microstate, even describe the dynamics that govern the K is going to change from one microstate to another. In addition, it would be quite useless to complete that monumental task. Even classically, if the position and momentum is given, there are 10^25 particles. There are lots of microscopic useless information which may be coarse-grained in order to determine a useful macroscopic property.

Equilibrium, being a thermodynamic property, only applies to macroscopic systems and its assumption follow to this coarse-grain.

There are requirements such as the scales of length and energy of K are much larger that those of their microscopic degrees of freedom. We define equilibrium of K with itself the system that satisfies the following conditions.

The system K is isolated The first assumption requires that the system under consideration do not dissipate energy in a larger environment. If we want to provide a system in equilibrium with an environment, K should be thought to include the system and the environment, in such way that the total system is isolated. This allows us to define a total energy E for K, that we assume to be much larger than the energy of it is the microscopic degrees of freedom.

All accessible microstates of K have the same probability The second assumption is where the coarse-graining occurs. We ignore all the complex and necessarily chaotic dynamics of the system in favor of a probability distribution for K that will be in any of the available microstates. We declare that this probability distribution is completely flat. Our goal is to deal with systems where this assumption fails obviously, so there is no major concern about the arbitrary assumptions. These two assumptions massively reduced the computational complexity of the problem and allowed us to calculate macroscopic observables.

10.1.3 Statistical Particles in Equilibrium

10.1.3.1 The Gibbs Distribution

Equilibrium assumptions could be use to determine the distribution of excitations in the balance of energy. This is possible because we assume the phase space density flat, as stated in the previous section.

Base on the two assumptions, to calculate the distribution of energy in a small subsystem of K can be achieved dividing K into a system S, a set of degrees of freedom that are of interest to us, and an environment R. Suppose that S has much smaller scales of length and energy than the K, and we do not pay attention to the microscopic states of R. The goal is to find the probability p_n that the K system is in a state such that S is a well-defined quantum microstate of energy ϵ_n. That is, we seek the probability that the total system is in an unknown state but with the system in a microstate that is known.

From the second hypothesis of the previous section we know that the probability of each microstate of the overall system K is equally probable. That means that if we want to find the probability that S is in a defined energy ϵ_n, we need only the number of States of K such that fill this condition. Thus, we have,

$$p_n = \Omega(\varepsilon_n) \tag{10.1}$$

where $\Omega(\epsilon_n)$, is called the statistical weight, and is the number of microstates of the system total such that S has energy ϵ_n. If we look at S to have energy ϵ_n, then we know that the overall system is in one of these states, but do not know which one. Of course, K is a macroscopic system, the number of its possible microstates is unimaginably huge. We will make the problem more manageable representing as an exponential,

$$p_n = e^{\ln \Omega(\varepsilon_n)} = \frac{1}{Z} e^{S(\varepsilon_n)} \tag{10.2}$$

where we have introduced the function $S(\epsilon_n) = \ln\Omega(\epsilon_n)$ for simplicity. This function is what we call the entropy. The logarithm of the number of microstates of K such that S has energy ϵ_n.

However, the first equilibrium assumption indicates that K is an isolated system, and as such, has a fixed total energy E. Because we know that the energy ϵ_n and the energy of the reservoir must be added to the total E, we can write $\epsilon_n = E(E - \epsilon_n)$ and let the entropy be a function of the energy of reservoir $E\epsilon_n$,

$$S(\varepsilon_n) \quad \to \quad S(E - \varepsilon_n) \tag{10.3}$$

Finally, because S had much smaller energy than K, we know that, we know that $E \gg \epsilon_n$, we are therefore justified to expand of the entropy in the exponential of Eq. (10.2). We have,

$$p_n \approx e^{S(E) - \varepsilon_n \left(\frac{\partial S}{\partial x} \right)_E} \tag{10.4}$$

Both $S(E)$, as well as the derivative of the entropy function in the Eq. (10.4) are constant with respect to ϵ_n. We can identify the derivative with a constant called "temperature,"

$$\beta = \frac{1}{T} = \left(\frac{\partial S}{\partial x} \right)_E \tag{10.5}$$

Of course, we also have to normalize the probability. Define a partition function $Z = \sum_n p_n = \sum_n p_n e^{\epsilon_n/T}$, we are left with the Gibbs distribution.

$$p_n = \frac{1}{Z} e^{-\beta \varepsilon_n} \tag{10.6}$$

These simple assumptions and abstractions allow to describe the states of extremely complex systems in a very simple and straightforward manner. In particular, we have never said whether the microscopic States of S were many body or single-particle states.

10.1.3.2 Bosons and Fermions

Consider a situation in which the overall system K is a collection of noninteracting quantum particles. It is assumed that the system is in equilibrium with itself at a temperature T and volume V. We must choose a small region of the total volume, if we look at this small volume, particles will come in and go out all the time. The total number of particles within this smaller volume will fluctuate due to interactions with the larger system at T. However, the number of particles should not fluctuate far from any average density for the system. We can model this type of fluctuation with a chemical potential μ. That is, we are phenomenologically describing fluctuations so that the number of particles N change but at a cost of energy μN in the presence of N particles.

Within the system, these N particles will be distributed between the quantum states k with energy ϵ_k. If there are n_k particles in the quantum state ϵ_k, then the contribution of the total energy is simply $n_k \epsilon_k$. The total energy of microstate of the system S is simply the sum of these energies over all these states. Thus, using the Gibbs distribution formula (10.6), the probability of that system S has N particles distributed with n_k particles in each state is given by,

$$p(N, \{n_i\}) = e^{\beta \mu N} \exp \left(\sum_i n_i \varepsilon_i \right) \tag{10.7}$$

with the condition that $\sum_k n_k = N$. It is possible that particles are distributed in quantum states in any way that satisfies this condition. Therefore, if we want to find the partition function for S, we have to add all of these possible configurations, as well as all possible particles in S. The partition function can be written as the sum over N and k of the Eq. (10.7),

$$Z = \sum_{N=0}^{\infty} \sum_{\{n_i\}} \exp \left[-\beta \sum_i (\varepsilon_i - \mu) n_i \right] \tag{10.8}$$

where for each value of N, we sum over all possible placements of N particles in quantum states k. The only condition is that the sum of the numbers of particles in each quantum state should add to the number of particles in the overall system, and therefore $\sum_k n_k \mu = N \mu$.

We know that all occupations n_k of each state must add to N, but we are adding on every possible value of N. Thus we can ignore the index N and simply sum over n_k. If we do this, the sums of the expression for the partition function reduced to,

$$Z = \sum_{n_1, n_2, \dots} \exp[\beta(\varepsilon_1 - \mu)n_1 - \beta(\varepsilon_2 - \mu)n_2 - \cdots] \qquad (10.9)$$

and we can factorize this sum. The occupation numbers are treated in the same footing so we write n instead of n_k, so the partition function can easily be written as,

$$Z = \prod_i \left\{ \sum_n e^{-\beta(\varepsilon_i - \mu)n} \right\} \qquad (10.10)$$

We know that quantum mechanically, the particles can be divided into bosons and fermions. Bosons have the property that many particles can be in the same quantum state, while the fermions are allow only a single Fermion state each. In terms of our partition function, this means that the sum over n in the Eq. (10.10) goes from $n = 0$ to $n = \infty$ for bosons while only from $n = 0$ to $n = 1$ for fermions. In the case of bosons, the sum in Eq. (10.10) is a geometric series which converges only if $e^{-\beta(\epsilon\mu)n} < 1$. Therefore, we know that the chemical potential μ should be negative for our analysis to be valid. If this is the case, we have for the bosonic partition function,

$$Z_B = \prod_i \left(1 - e^{-\beta(\varepsilon_i - \mu)}\right)^{-1} \qquad (10.11)$$

Using the fact that the number of average occupation number can be determined through the formula $\angle n_k \rangle = -\frac{1}{\beta} \frac{\partial}{\partial \epsilon_k} Ln Z$, we find that the distribution of bosonic occupations between in the quantum states k is given by,

$$\langle n_i \rangle = \frac{1}{e^{\beta(\varepsilon_i - \mu)} - 1} \qquad (10.12)$$

Now let's look at the case we are dealing with fermions. No two fermions can be in the same quantum state, n can be equal to 0 or 1 in the Eq. (10.10). Our partition function again has a simple form,

$$Z_F = \prod_i \left(1 + e^{-\beta(\varepsilon_i - \mu)}\right) \tag{10.13}$$

and again we can use it to find the Fermionic occupation statistics,

$$\langle n_i \rangle = \frac{1}{e^{\beta(\varepsilon_i - \mu)} + 1} \tag{10.14}$$

The sum over N to infinity in the Eq. (10.8) contradicts the assertion that S is very small in comparison with K on what the legitimacy of the Gibbs distribution depends. This is not usually a problem because the probability of high N states will be very low, they are exponentially suppressed in the energy of the States. However, this is not the case when phenomenon like Bose-Einstein condensates are considered. In other areas, it is especially useful in the theory of solids at low temperatures, where these functions of equilibrium distribution play a major role. Even when interactions are included, they still retain some validity, as starting points from which to perturbative expansions are made around weak coupling. It is often possible to reformulate a problem of interaction in terms of elementary collective excitations which are non-interacting, with bosonic or Fermionic statistics.

10.2 SYSTEMS OUT OF EQUILIBRIUM

A non-equilibrium system has no well-defined thermodynamic properties. Entropy, temperature, and energy-free are undefined for a general system. In another sense, however, the loss of equilibrium quantities can be something of a blessing which obliges to move away from abstract thermodynamic concepts available only in non-realistic limits and limit us to more specific observable such as Ensemble averages and correlation functions.

10.2.1 Observables Out of Equilibrium

From an intuitive perspective, we fall into two basic categories: the set of average operators of a single particle and correlations or averages of all the products of such operators. These are the quantities that are directly measured in experiments. In equilibrium, the knowledge of these specific observable thermodynamic quantities such as temperature is inferred. Equilibrium will be sufficient to know only a concrete observable.

We made use of two basic recipes. The first is given in terms of density matrix formalism, describing instantaneous ensembles averages through the relation,

$$\langle A \rangle = \text{Tr}\left[\rho A\right] \tag{10.15}$$

When the system is not in equilibrium, $\hat{\rho}$ usually will be a function of time. As a result, the averages ensemble will be time-dependent through traces taken over $\hat{\rho}$. When we considered only unitary dynamics, these are governed by the equation by Von Neumann, but we know that the inclusion of an environment with which the system has not equilibrium will cause a non-unitary dynamics. We have to introduce the quantum master equation.

The second recipe for the solution of problems of non-equilibrium is expressed in terms of correlation functions. Correlations are also experimentally observable in the statistical properties of the operators of the single particle distributions. In addition, if we have a separation in energy scales between the terms that contribute to the total Hamiltonian, then the correlation functions have well understood expansions in terms of Feynman diagrams. Far from equilibrium, our understanding of these functions must be generalized from what we know in equilibrium or unitary dynamics. For example, functions of non equilibrium depends on $G_{x,x'}(t,t') = \langle a_x^\dagger(t)a_{x'}(t)\rangle$ depend on t and t' separately instead of the difference tt' as in is the case of equilibrium. The brackets $\langle ... \rangle$ stand as average on density matrix that are time-dependent instead of stationary thermal ensembles.

10.2.1.1 *Non-Equilibrium Correlation Functions*

The averages of products of operators is written in general as follows,

$$G_{\alpha,\alpha'}(t,t') = -i\langle Ta_\alpha^\dagger(t)a_{\alpha'}(t')\rangle \tag{10.16}$$

where the brackets $\langle \rangle$ denotes an average over the ground quantum state of the full Hamiltonian and T_D is the symbol for the time-order operator. At $T = 0$, we can formally obtain this average and write the correlation function in terms of a dispersion or scattering matrix without ever knowing the ground state.

We begin to remember the recipe in equilibrium with a bath at temperature zero. We write the total Hamiltonian as $H = H_0 + V$ where H_0 is the bare Hamiltonian with a known ground state $|\psi_0\rangle$. We assume that we can diagonalize V, so we want to include its effect as a perturbation. Being in equilibrium, the Hamiltonian is supposed to be static. We will assume that in the infinite past, $V = 0$. As time progresses, V adiabatic is turned on so that the system is instantly in the state $H_0 + V$ for any future time. Then, in the infinite future, the perturbation is adiabatically switched off again. Assuming the perturbation to be fully on in $t = 0$, then we can write the instantaneous ground state $|\psi\rangle$ over which the average in Eq. (10.16) is written as,

$$|\phi\rangle = S(0, -\infty)|\phi_0\rangle \tag{10.17}$$

$$\langle\phi| = \langle\phi_0|S(-\infty, 0) \tag{10.18}$$

in the interaction picture with respect to V. The operator $S(t, t') = U(t)U^\dagger(t')$ is the scattering operator. If the time evolution of a state is defined such that $|\psi(t)\rangle = U(t)|\psi(0)\rangle$ then the operator $S(t, t')$ acts in a more general way as,

$$|\psi(t)\rangle = S(t, t')|\psi(t')\rangle.$$

This operator can be written as a time ordered operator,

$$S(t, t') = T \exp\left[-i \int_{t'}^{t} dt_1 V(t_1)\right] \tag{10.19}$$

Because we turned on and off the perturbation, the adiabatic theorem says that the state at $t = \infty$ is equal to the State at $t = \infty$ up to a phase, this means,

$$\langle\phi_0|S(\infty, -\infty)|\phi_0\rangle = e^{iL} \tag{10.20}$$

This result is the fundamental to formally relax when, in order to generalize this out of equilibrium. Far from equilibrium, the perturbation cannot be activated adiabatic. There is no reason to believe that the system will remain in their ground state in the infinite future. Using the Eq. (10.20), we can write $\langle\psi|$ in terms of the ground state at the infinite future instead of the infinite past, up to the phase,

$$\langle\phi| = \frac{\langle\phi_0|S(\infty, 0)}{\langle\phi_0|S(\infty, -\infty)|\phi_0\rangle} = e^{-iL}\langle\phi_0|S(\infty, 0) \tag{10.21}$$

The end result of all these mathematical gymnastics is that we can now write the correlation function in the interaction picture as an time-ordered product of the operators of the particle and the matrix $S(\infty, \infty)$, averaged over the ground state $|\psi_0\rangle$ of the bare Hamiltonian,

$$G_{\alpha,\alpha'}(t, t') = -i\frac{\langle Ta_\alpha^\dagger(t)a_{\alpha'}(t')S(\infty, -\infty)\rangle_0}{\langle\phi_0|TS(\infty, -\infty)|\phi_0\rangle} \qquad (10.22)$$

Thus, Eq. (10.22) isn't useful, but when combined with the Eq. (10.19) and equation,

$$U(t) = 1 + \sum_{n=1}^{\infty}\left(\frac{-i}{\hbar}\right)\int_0^t dt_1 \int_0^{t_1} dt_2 ... \int_0^{t_{n-1}} dt_n H(t_1)H(t_2)...H(t_n),$$

$$(10.23)$$

is a series expansion for G in the small Hamiltonian V.

To generalize the Eq. (10.22) to non-equilibrium situations, we go back to the assumption of equilibrium. The adiabatic behavior that turns on and off the perturbation very slowly compared to the system time allowed us to write the ground state in the infinite future in terms of the state in the infinite past and an additional phase in Eq. (10.21). If this assumption is relaxed, we neglect Eq. (10.21) and return to the Eq. (10.18), written in slightly different form using the transitivity of the S operator,

$$\langle\phi| = \langle\phi_0|S(-\infty, 0) = \langle\phi_0|S(-\infty, \infty)S(\infty, 0) \qquad (10.24)$$

We simply write the future state as the infinite past state evolved to $t = \infty$, i.e., a phase. This method is equally valid far from equilibrium. Substituting Eq. (10.24) in the expression for the correlation function, we find that we can write it as,

$$G_{\alpha,\alpha'}(t, t') = -i\langle T_C a_\alpha^\dagger(t)a_{\alpha'}(t')S(-\infty, -\infty)\rangle_0 \qquad (10.25)$$

where C is a contour in the complex time between $t = \infty$ to ∞ slightly "above" the real axis t from $t = \infty$ to ∞ slightly "below" the real axis.

The forward-backward evolution is included in a single scattering operator which is written in terms of this Keldysh contour C [4]. Backward evolution occurred after the forward evolution, the operator T_C orders operators in time along this contour C. The scattering matrix now has the form,

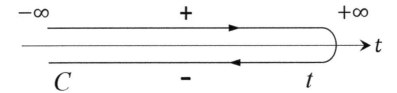

Figure 10.1: Keldysh contour.

$$S(-\infty, -\infty) = T_C \exp\left[-i \int_C dt_1 V(t_1)\right] \tag{10.26}$$

where we use $\hbar = 1$. However, since the arguments of time can be on different paths of the contour, we need to reorganize the Dyson equation as a matrix.

This means that if t is in the forward path while t' is in the backwards path, then the symbol of contour ordered T_C evaluates in Eq. (10.25) as,

$$G_{\alpha,\alpha'}(t,t') = G^>_{\alpha,\alpha'}(t,t') = -i\langle a_{\alpha'}(t')a_\alpha^\dagger(t)S(-\infty,-\infty)\rangle_0 \tag{10.27}$$

regardless of the real values of t and t'.

If t and t' are in the forward path, then T_C becomes the time order operator T_D. There are four possible ways that t and t' can be placed in the two paths of the contour C. Thus, there are four possible Green's functions that can reach the correlator. Our goal is to expand the total correlator in (10.25) in terms of bare correlators $g_{\alpha,\alpha'}(t,t') = \langle T_C \alpha'_\alpha(t')a_\alpha^\dagger(t)\rangle$. This expansion is best achieved through the bare correlators corresponding to the four orderings by T_C. For bosons it is,

$$g^>_{\alpha,\alpha'}(t,t') = -i\langle a_{\alpha'}(t')a_\alpha^\dagger(t)\rangle_0 \tag{10.28}$$

$$g^<_{\alpha,\alpha'}(t,t') = -i\langle a_\alpha^\dagger(t)a_{\alpha'}(t')\rangle_0 \tag{10.29}$$

$$g^T_{\alpha,\alpha'}(t,t') = \theta(t-t')G^>_{\alpha,\alpha'}(t,t') + \theta(t'-t)G^<_{\alpha,\alpha'}(t,t') \tag{10.30}$$

$$g^{\tilde{T}}_{\alpha,\alpha'}(t,t') = \theta(t'-t)G^>_{\alpha,\alpha'}(t,t') + \theta(t-t')G^<_{\alpha,\alpha'}(t,t') \tag{10.31}$$

With the expansion of the matrix of scattering in the Eq. (10.25), have for the lesser correlator,

$$G^<_{\alpha,\alpha'}(t,t') = -i\sum_n \frac{(-i)^n}{n!} \int_C dt_1 \dots dt_n \langle T_C V(t_1) \dots V(t_n)a_\alpha^\dagger(t)a_{\alpha'}(t')\rangle_0$$

$$\tag{10.32}$$

To illustrate how we can represent this series very simply in terms of matrix multiplication, let propose in the Schrödinger picture, an interaction with the form,

$$V = \sum_{\lambda,\beta} M_{\lambda\beta} a_\lambda^\dagger a_\beta \tag{10.33}$$

Equation (10.32) then becomes,

$$G_{\alpha,\alpha'}^<(t,t') = g_{\alpha,\alpha'}^<(t,t')$$
$$+ \sum_{\lambda\beta} M_{\lambda\beta} \int_C dt_1 \langle T_C a_\alpha^\dagger(t) a_\beta(t_1) \rangle_0 \langle T_C a_\lambda^\dagger(t_1) a_{\alpha'}(t') \rangle_0 + \cdot \tag{10.34}$$

Remember that the integral in t_1 runs along the contour C, we know that T_C will order the operators on the R.H.S. depending on which path is t_1. The are two possibilities. That t_1 is in the forward path and we have $\langle T_C a_\alpha(t) a_\beta^\dagger(t_1) \rangle_0 \rightarrow g^<$ and $\langle T_C a_\lambda(t_1) a_{\alpha'}^\dagger(t') \rangle_0 \rightarrow g^T$. If t_1 is on the backward path we have $\langle T_C a_\lambda(t) a_\beta^\dagger(t_1) \rangle_0 \rightarrow g^{\tilde{T}}$ and $\langle T_C a_\lambda(t_1) a_{\alpha'}^\dagger(t') \rangle_0 \rightarrow g^<$.

Because $t1$ will be on both paths, we can write these two possibilities in the same line as,

$$G_{\alpha,\alpha'}^<(t,t') = g_{\alpha,\alpha'}^<(t,t')$$
$$+ \sum_{\lambda\beta} M_{\lambda\beta} \int_{-\infty}^{+\infty} dt_1 [g_{\alpha,\beta}^<(t,t_1) g_{\lambda,\alpha'}^T(t_1,t') - g_{\alpha,\beta}^{\tilde{T}}(t,t_1) g_{\lambda,\alpha'}^<(t_1,t')]$$
$$+ \cdots \tag{10.35}$$

If we define the following matrix,

$$\bar{G}_{\alpha,\alpha'}(t,t') = \begin{pmatrix} G_{\alpha,\alpha'}^T(t,t') - G_{\alpha,\alpha'}^>(t,t') \\ G_{\alpha,\alpha'}^<(t,t') - G_{\alpha,\alpha'}^{\tilde{T}}(t,t') \end{pmatrix} \tag{10.36}$$

then, we see immediately that the information contained in the Eq. (10.35) is collected in the equation,

$$\bar{G}_{\alpha,\alpha'}(t,t') = \bar{g}_{\alpha,\alpha'}(t,t') + \sum_{\lambda\beta} M_{\lambda\beta} \int_{-\infty}^{+\infty} dt_1 \bar{g}_{\alpha,\beta}(t,t_1) \bar{g}_{\lambda,\alpha'}(t,t') + \cdots$$

where g is the definition of matrix similar to the Eq. (10.36) for the functions $g_{\alpha\alpha'}(t, t')$. Relation (10.37) holds even when the brackets $\langle...\rangle$ indicates an average over a mixed state, rather than merely the average of the ground quantum state.

10.2.2 Equations of Motion

Having considered the types of observables of interest in the theory of non-equilibrium, we have to talk about the equations of motion governing them. The density operator $\rho(t)$ is a dynamical measure of the population of each state of quantum particles. When we speak about unitary dynamics, we observe that the density operator solves the equation of Von Neumann,

$$i\hbar\frac{\partial\rho}{\partial t} = [H, \rho]$$ (10.37)

To include the effects of an environment we define the total Hamiltonian,

$$H = H_S + H_R + V$$ (10.38)

The Hamiltonian that only affects the system is designated by H_S. The Hamiltonian H_R regulates the dynamics of the reservoir. The link between the system and the reservoir is given by V, which we assume to be weak. This hypothesis is required by construction. By assuming there is a partition of the overall system in a system and a reservoir, implicitly we assume that the link between these partitions is weak enough for the dynamics of the two can be separated. The density matrix that describes the state of the system + reservoir will continues to obey the Von Neumann equation, with the total Hamiltonian given in Eq. (10.38). The dynamics of this total system is unitary. However, remember that we are not interested in the dynamics of the overall system. We are only interested in S. We want to coarse grain on all the unnecessary information about the reservoir to obtain a equation of motion only for S.

We can define the reduced density matrix $\sigma(t)$ so that the system,

$$\sigma(t) = \text{Tr}[\rho]$$ (10.39)

Then we pass to the interaction picture,

$$\bar{\rho}(t) = e^{i(H_S+H_R)t/\hbar} \rho(t) e^{-i(H_S+H_R)t/\hbar} \tag{10.40}$$

$$\bar{V}(t) = e^{i(H_S+H_R)t/\hbar} V e^{-i(H_S+H_R)t/\hbar} \tag{10.41}$$

Based on this, the Von Neumann equation only details the dynamic dependency on the coupling term,

$$i\hbar \frac{\partial \bar{\rho}}{\partial t}(t) = [\bar{V}(t), \bar{\rho}(t)] \tag{10.42}$$

The interaction picture is particularly useful due to our hypothesis that the coupling is weak. Because we have assumed that V is small compared to other scales of energy of the problem, the time evolution given by the Eq. (10.42) will be slow without rapid evolution due to other terms in the Hamiltonian. We can integrate both sides of the Eq. (10.42) to obtain,

$$\bar{\rho}(t + \Delta t) - \bar{\rho}(t) = \frac{1}{i\hbar} \int_t^{t+\Delta t} dt' [\bar{V}(t'), \bar{\rho}(t')] \tag{10.43}$$

We can iterate this equation by letting $\tilde{\rho}(t') = \tilde{\rho}(t + \Delta t)$ and using the Eq. (10.42) again. This can be done in any desired order in $\tilde{V}$. To second order, we have,

$$\bar{\rho}(t + \Delta t) = \bar{\rho}(t) + \frac{1}{i\hbar} \int_t^{t+\Delta t} dt' [\bar{V}(t'), \bar{\rho}(t')]$$
$$+ \left(\frac{1}{i\hbar}\right)^2 \int_t^{t+\Delta t} dt' \int_t^{t'} dt'' [\bar{V}(t'), [\bar{V}(t''), \bar{\rho}(t'')]] \tag{10.44}$$

The Eq. (10.44) is exact, but to get something useful from it, we must assume certain properties of the reservoir. First we must say that the reservoir is large enough as coupling with the system does not produce any change on it,

$$\bar{\sigma}_R(t) = \text{Tr}_S[\bar{\rho}(t)] = \sigma_R \tag{10.45}$$

where σ_R is not a function of time. Second, we assume that the reservoir is in a stationary state. Its density operator commutes with the Hamiltonian.

$$[\sigma_R, H_R] = 0 \tag{10.46}$$

Thus, σ_R is diagonal in the energy base, and we can consider that is mixed state of energy eigenstates. Finally, we will make a strong hypotheses about the coupling. We will assume that V is a product of a system observable A consisting of operators that act only on the states of the system and an observable of the reservoir R. By construction, these operators commute among themselves and we have,

$$V = -AR \qquad (10.47)$$

where is easy to generalize to a sum of such operators. We also assume that the average value of R vanishes. The trace of R in interaction picture is,

$$\text{Tr}[\sigma_R \bar{R}(t)] = 0 \qquad (10.48)$$

Looking at the reservoir correlation function $g(t, t') = \langle R(t)R(t') \rangle$ is easy to demonstrate that this function depends only on the difference $\tau = tt'$ virtue of the fact that the reservoir is in a stationary state. We have

$$\begin{aligned}
g(t, t') &= \text{Tr}[\sigma_R \bar{R}(t)\bar{R}(t')] \\
&= \text{Tr}[\sigma_R e^{iH_R t/\hbar} \bar{R} e^{-iH_R(t-t')/\hbar} \bar{R} e^{-iH_R t'/\hbar}] \\
&= \text{Tr}_R[\sigma_R \bar{R}(\tau)\bar{R}(0)] \\
&= g(\tau) \qquad (10.49)
\end{aligned}$$

This means essentially that the reservoir has a very dense ensemble of energy levels. Thus, when τ grows enough, the exponents in the expansion of the Fourier transform of $g(\tau)$ interfere destructively because of the frequencies in the expansion are large spread and is very dense. Therefore, $g(\tau) \to 0$ for times much larger than some correlation of bath time τ_C. The assumption is hat τ_c is very small in comparison with the relevant time scales of the problem.

Under these assumptions, our goal is to use the exact Eq. (10.44) to determine an equation of motion for the density system operator $\tilde{\sigma}(t)$ where the degrees of freedom in the reservoir have been averaged. We start writing our total density matrix as the sum of two parts,

$$\bar{\rho}(t) = \text{Tr}_R\bar{\rho}(t) \otimes \text{Tr}_S\bar{\rho}(t) + \bar{\rho}_{\text{correl}}(t) \qquad (10.50)$$

We have written the density matrix as the product of the two reduced matrices plus a contribution, related to the correlations between the system and bath. We will see that the assumption allows us to neglect the contribution of $\tilde{\rho}_{cor}$. Back to the Eq. (10.44), assume that Δt is small with respect to the time evolution of the system τ_S. If this is the case, then we can replace $\tilde{\rho}(t'')$ with $\tilde{\rho}(t)$. We have ignored the expansion of the system to third order and higher terms in V, arguing that a substantial change in the system under these conditions takes much more time than thee one to be considered. We can trace both sides of this equation over the bath degrees of freedom. Because the second term in the Eq. (10.44) vanishes due to (10.48), we have,

$$\frac{\Delta \bar{\sigma}}{\Delta t} = -\frac{1}{\hbar^2 \Delta t} \int_t^{t+\Delta t} dt' \int_t^{t'} dt''$$
$$\text{Tr}_R[\bar{V}(t'), [\bar{V}(t''), \bar{\sigma}(t) \otimes \sigma_R + \bar{\rho}_{\text{correl}}(t)]] \qquad (10.51)$$

where we have divided both sides by Δt to form an equation that recalls a time derivative.

An estimate of the order of magnitude of the term due to correlations of the bath system including a contribution of the linear term, is given as

$$\left(\frac{\bar{\sigma}}{\Delta t}\right)_{\text{correl}} = -\frac{1}{\hbar^2 \Delta t} \int_{-\infty}^{t} dt'' \int_t^{t+\Delta t} dt' \langle \bar{V}(t'') \bar{V}(t') \rangle_R \qquad (10.52)$$

The bath correlator vanishes for $t' - t \gg \tau_S$, an estimation of this quantity is given by,

$$\frac{v^2 \tau_c^2}{\hbar^2 \Delta t} = \frac{\tau_c}{T_S \Delta t} \qquad (10.53)$$

where is v is the energy that characterizes the strength of V. Thus, in the limit where $\frac{\tau_C}{T_S \Delta t} \to 0$, we can write the equation of motion of coarse grained system for our density matrix $\tilde{\sigma}$ in the interaction picture with respect to V as,

$$\frac{\Delta \bar{\sigma}}{\Delta t} = -\frac{1}{\hbar^2 \Delta t} \int_t^{t+\Delta t} dt' \int_t^{t'} dt''$$
$$\text{Tr}_R[\bar{V}(t'), [\bar{V}(t''), \bar{\sigma}(t) \otimes \sigma_R]] \qquad (10.54)$$

A better way to write this is in term of the bath correlations functions, that are function of $\tau = t' - t''$. Let's replace our integration variables such that time integrations can be written as,

$$\int_t^{t+\Delta t} dt' \int_t^{t'} dt'' = \int_0^{\Delta t} d\tau \int_{t+\tau}^{t+\Delta t} dt' \qquad (10.55)$$

Recalling that the correlators of the reservoir vanish for $\tau \ll \tau_C$, we know that we can extend the upper limit of integration of $d\tau$ to infinity and lower limit of integration to dt' to t. Finally, we can expand the commutator in Eq. (10.56) to obtain,

$$\frac{\Delta \bar{\sigma}}{\Delta t} = -\frac{1}{\hbar^2} \int_0^{+\infty} \frac{1}{\Delta t} \int_t^{t+\Delta t} dt'$$
$$\times \left\{ g(\tau)[\bar{A}(t')\bar{A}(t'-t)\bar{\sigma}(t) - \bar{A}(t'-\tau)\bar{\sigma}(t)\bar{A}(t')] \right.$$
$$\left. + g(-\tau)[\bar{\sigma}(t)\bar{A}(t'-\tau)\bar{A}(t') - \bar{A}(t')\bar{\sigma}(t)\bar{A}(t'-\tau)] \right\} \quad (10.56)$$

where the operators A are representation of the interaction picture of the system in Eq. (10.47). Equation (10.56) is the quantum master equation. It is an operator equation, so it can be written in any base we choose. The products of the operators appearing in the Eq. (10.56) can be organized into what are called Lindblad operators. Often, we can identified this terms as noise and dissipation.

CHAPTER 11

INTRODUCTORY SPINTRONICS

11.1 ABOUT SPIN

Conventional electronics uses the electron charge flow as the basic component of the Boolean construction of states, but electrons has an intrinsic angular moment, called spin. Spin, denoted with vector $\vec{S}$, is an intrinsic property, characterized by quantum number $1/2$: in fact, all protons, neutrons, and electrons have spin $s = 1/2$. This leads to a total angular moment,

$$S = \sqrt{\frac{1}{2}\left(\frac{1}{2}+1\right)}\,\hbar = \frac{\sqrt{3}}{2}\hbar \qquad (11.1)$$

$\vec{S}$ cannot be measured simultaneously on the three spatial directions with arbitrary precision, due to the Indeterminacy Principle, but we can measure the components along the axes, which are also quantized and where there are only two orientations of the z component.

This implies that they can only orient themselves in the modes corresponding to projections equal to $+\hbar/2$ and $-\hbar/2$, as shown in the figure. Commonly called spin-ups and spin-downs.

There are two experiments that justify the presence of the spin: the division of the hydrogen spectrum lines and the Stern-Gerlach experiment. The Stern-Gerlach, in 1922 [126], showed that a bundle of silver atoms passing through an increasing intensity magnetic field divides into two beams: all

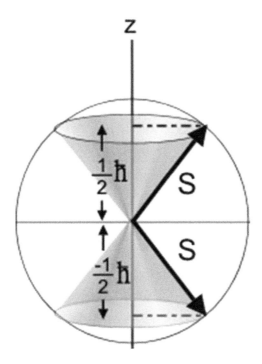

Figure 11.1: Spin representation with spatial quantization.

this leads to the proof that a single electron has an intrinsic angular moment and a magnetic moment. Classically this is resemble if we think of the electron as a sphere that rotates on itself: Although the spin is not rotating, is from this consideration that the spin name use.

The electron therefore has an intrinsic angular moment, which is independent of the orbital angular momentum. At every angular moment, a magnetic moment is always associated: this implies that electrons, interacting with a magnetic field, behave as tiny magnetic needles. The z component of the associated magnetic moment is expected to be,

$$S_z = m_s \hbar, \quad \text{with} \quad m_s = \pm\frac{1}{2} \tag{11.2}$$

as the magnetic moment associated with the orbital component of the angular moment, but experimentally have the values, where e and m_e are charge and mass of the electron respectively, and g is called the gyromagnetic constant: for the electron this is $g = 2.00232$. The exact value of g

is predicted by quantum mechanics in the Dirac equation [114]. From this comes a natural constant called Bohr Magneton, defined as,

$$\mu_z = \pm \frac{1}{2} \left(\frac{e\hbar}{2m_e} \right) \qquad (11.3)$$

The spin magnetic moment is important in spin-orbit interactions and interactions between atoms and an external magnetic fields.

The term spin should not be taken literally, that is, in the classical sense of a rotating sphere on itself. A rotating sphere around its axis produces a magnetic moment, but the intensity of this must be reasonable in relation to the size of the sphere itself. High energy diffusion experiments show that the electron does not have "body" up to a resolution of 10^{-3} fermi: at these levels, to produce the observed magnetic moment, a rotation of about $10^{32} rad/s$ is needed, too excessive even for atomic world.

11.1.1 Importance of Spin

There are four important reasons to introduce spin. First, the connection between spin and magnetism, necessary for storing information. The spin is connected to ferromagnetic materials because spontaneous magnetization allows the electronic states to become spin-dependent, while for non-magnetic materials the electronic states come in pairs with the same energy but with opposite spin, which leads to a density states independent of spin. Basically, in a ferromagnetic material, the density of states is different for spin-up and spin-down. Since many transport properties depend on the density of states near the Fermi energy, asymmetry in the density of states allows ferromagnets to generate, manipulate, and reveal spin. In addition, ferromagnetic materials possess the property of hysteresis: if the ferromagnet is subject to a magnetic field, a history of a cycle in applying and removing the magnetic field remains, called hysteresis.

Second, the magnetic properties vary greatly depending on the chemical composition, but also by thermal treatments or mechanical stresses. In terms of memory, the two zero magnetic field stable states correspond to the logic levels 0 and 1 or a bit. Data can be written by applying a magnetic field to align the magnetization, creating a stable state when the magnetic field is turned off: ferromagnetics are therefore the natural candidates for information storage. There is a connection between spintronics and the storage of data.

Third, the spin is connected to quantum mechanics. In classical mechanic the angular momentum can be divided into two parts, an orbital angular momentum and a spin angular momentum. If one considers Earth's motion around the Sun, this generates orbital angular momentum, while rotation generates angular momentum of spin. In an atom, the motion is governed by quantum mechanics, and the idea of spin was born when Dirac combined quantum mechanics with special relativity in the 1920s [127]. By solving Dirac's equation, one of the consequences is the need for an internal property, the spin. Given its quantum mechanical origin, it enjoys particular properties: its value along any axis can always take only two values, $S_z = \pm\hbar/2$. It also obeys the principle of Heisenberg's indetermination, that is, the three components of the spin (S_x, S_y, S_z) cannot be measured arbitrarily at the same time.

From the point of view of computation, it is more important that spin can be in a quantum superposition of states, $\alpha|\uparrow\rangle + \beta|\downarrow\rangle$, where α and β are complex numbers. Here we can define a qubit: an ordinary bit can assume two values, 0 or 1, while the qubit, in addition to the values 0 and 1, may be in an overlapping state $\alpha|\uparrow\rangle + \beta|\downarrow\rangle$, where $|\alpha|^2$ is the probability of finding it up and $|\beta|^2$ the probability of finding it down. Qubit is the basis for a new paradigm of computing, called quantum computing.

Finally, the ever-increasing device size reduction, the role of the spin becomes more and more important. Understand how spin and magnetism interact on a nanometer scale is an area of interest for the development of spintronics.

Spintronics has the advantage to achieve performance by increasing speeds with low power consumption. As the dimensions decrease, the problems of power dissipation and the temperature are becoming more and more important. Spin brings some power benefits: a ferromagnetic bit can store information without power consumption to keep the data. In addition, spin can bring benefits also in terms of speed: in electronics the velocity is determined by the constant RC. Using spin, it should be possible to overcome this general rule. For example, precession of spin or magnetism is not governed by the constant RC. Therefore, there is potential for greater speeds and low power applications, but new circuit architectures are needed to achieve these results, though a more thorough analysis of actual consumption must be performed.

11.2   EFFECTS ASSOCIATED TO SPINTRONICS

11.2.1   Interlayer Exchange Coupling

The most known property of ferromagnets is the magnetic force. In a multi-layer metal, composed of alternate ferromagnet (FM) states and a diamagnetic material (NM), a magnetic dipole coupling is created between the pairs of layers of FM, as well as between two neighboring magnets. When the distance between neighboring FM layers decreases, and becomes small enough, a new type of coupling is born, as a result of quantum mechanical exchanges. This phenomenon, discovered in 1986, takes the name of Interlayer Exchange Coupling.

The first theoretical work goes back to the 1950s, thanks to Raudermann, Kittel, Kasuya, and Yoshida (RKKY) [128]. Given two magnetic moments m_1 and m_2, within a diamagnetic metal, we consider what happens when a conducting electron passes from m_1, propagates and disperses to m_2. If the dispersion depends on the spin of the electron, then the energy of the system depends on the orientation of the magnetic moments and can be written as,

$$\mu_z = \pm \frac{1}{2} \left(g \frac{e\hbar}{2m_e} \right) \tag{11.4}$$

This coupling implies parallel moments for $J_{RKKY} > 0$, or ferromagnetic coupling, while anti-parallel moments for $J_{RKKY} < 0$, or antiferromagnetic coupling. The coefficient of the coupling is written,

$$\mu_B = \frac{e\hbar}{2m_e} \tag{11.5}$$

where k_F is the Fermi wave vector of the conducting electron, and r is the distance between the two magnetic moments. This result implies that the magnetic orientation alternates between parallel and antiparallel as a function of the distance, given the electron's wave nature. The RKKY theory was used to understand interactions in diluted magnetic alloys but was observed only in the 1980s with the advent of magnetic multilayer systems.

Due to the very narrow wavelength the width control on the layers is necessary, and it has only been achieved in the process called Molecular Beam Epitaxy (MBE).

The Interlayer Exchange Coupling (IEC) with antiferromagnetic coupling, $J < 0$ was observed in 1988 in Fe / Cr / Fe systems. The oscillatory nature of the IEC has been investigated in FM/NM/FM systems with NM material of different nature and width. Depending on the width of the NM layer, the FM layer magnetization oscillates between parallel and antiparallel alignments. In addition, the coupling force oscillates according to the width of the NM layer, as predicted by the RKKY theory. By applying the RKKY theory to an FM/NM/FM system, it is necessary to add the components of the FM / NM interactions, obtaining a factor,

$$E = -J_{RKKY}\mathbf{m}_1 \cdot \mathbf{m}_2 \qquad (11.6)$$

where d is the width of the NM layer. However, experimental data follows another curve, giving,

$$J_{RKKY} \sim \frac{\cos(2k_F r)}{r^3} \qquad (11.7)$$

where k_B is a wave vector due to the reticular structure of the material. This is because NM layer is not perfectly continuous, since it is formed by atoms that have a spacing that depends on the composition of the lattice. The difficulties in growing specimens of particular NM material thickness and in measuring thickness have increase studies on wedge-shaped samples.

While the RKKY model best describes the IEC dependence on the thickness of NM material, it does not explain IEC oscillations depending on the width of the FM layer. Several models based on quantum wells have been proposed, forming a more complete picture of the IEC mechanism complex than the RKKY model alone. The photoemission experiments have confirmed the importance of quantum wells in the IEC.

11.3 GIANT MAGNETIC RESISTANCE

The phenomenon of giant magnetoresistance was discovered in 1988, in a Fe / Cr system, where the thickness of the chromium layer is such that it has an IEC in anti-ferromagnetic coupling. The anti-parallel alignment was changed to parallel with the application of a sufficiently large external magnetic field. When the in-plane resistance was measured, there was a change of 50% as a function of the magnetic field. It was well known that the resistance of conductors changed value under the effect of a magnetic field, but with variations of the order of a few percentage points, and especially not for such small fields: given such a high variation, it was called "giant." It was immediately clear that the giant magneto resistance (GMR) could be use for storage of information.

The discovery of GMR shortly after IEC studies is not a casual fact: resistance to a structure that uses GMR is closely related to the magnetization of its layers. An electrical current through a metal always has a resistance R, for many reasons. In a crystal the atoms vibrate around their equilibrium position, slightly exited from the perfect equilibrium position in the lattice, and electrons may be diverted from these. Another important contribution is given by the dispersion of electrons against the impurities of the material. The electrons involved in electrical conduction are mainly those at Fermi's level. In paramagnetic materials there is no difference between spin-up or spin-down electrons, which also contribute to resistance. In a ferromagnetic material, electrical resistance is a slightly different phenomenon. Theoretical behavior of GMR and conduction can simply be understood by a resistor model. First, GMR is not due to interaction between conducting electrons and a magnetic field but between conducting electrons and FM layers. The role of the magnetic field is to change the alignment of the magnetization between adjacent non-parallel parallel FM layers and vice versa.

The basic idea is the difference in resistance between two conducting channels, one for spin-up electrons, and the other for spin-down electrons, as proposed by Mott in 1936 [129]. For simplicity, let us consider an FM/NM/FM system, and consider the resistances in the case of alignments of parallel and antiparallel magnetizations.

The current must be thought as two independent channels: a channel formed only by spin-up electrons and one by only spin-down electrons. Let us assume there is a spin-dependent dispersion in the FM layer, greater in the case of spin parallel to the magnetization, lesser than opposite. The dispersion, to which electrons are subjected depends on the density of

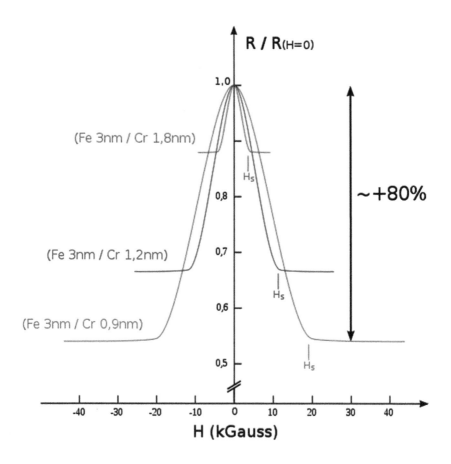

Figure 11.2: Giant magnetic resistance in the Fe/Cr system [The figure shows three Fe/Cr systems with different thicknesses of Cr. The electrical resistance changes by applying a few kG of magnetic field, up to a reduction of about 80% for Cr often 0.9 nm.].

states. The two spin channels are subject to two different degrees of dispersion, and macroscopically this leads to different resistances. Additionally, let us assume that the thickness of the layers is smaller than the average free electron path: this is important to make sure that a large number of electrons pass through both layers of the FM, despite the current flowing flat over the layers.

In the case of anti-parallel magnetization, consider the spin-up channel first: an electron will suffer a strong scatter in the left layer, and weak in the right layer. Here the dispersion is the source of resistance: the contribution of the left layer will give a great resistance, which we will call R_{large}, while the right layer will contribute to a smaller resistance, called R_{small}. Since, an electron suppose to cross both layers, the total resistance is the sum of the two, $R_{AP}^{\uparrow} = R_{large} + R_{small}$

Conversely, for a spin-down electron, a small resistance from the left and large layer of the right layer is obtained, thus obtaining the same final resistance, $R_{AP}^{\downarrow} = R_{large} + R_{small}$.

The total resistance of the anti-parallel alignment is given by the parallel resistances of the individual channels,

$$R_{AP} = \frac{R_{\text{small}} + R_{\text{large}}}{2} \qquad (11.8)$$

In the case of anti-parallel magnetization, a spin-up electron gives strong diffusion both in the left and right FM layers, leading to resistance, $R_P^{\uparrow} = R_{large} + R_{large}$. The spin-down electron, however, has weak diffusion in both layers of the FM, obtaining $R_P^{\downarrow} = R_{small} + R_{small}$. The total resistance of this configuration is still the parallel of the two channels:

$$R_P = \frac{2R_{\text{small}}R_{\text{large}}}{R_{\text{small}} + R_{\text{large}}} \qquad (11.9)$$

Taking the $R_{small} \to 0$ limit case, you immediately see $R_P \to 0$ while R_{AP} remains $R_{large}/2$. Since the resistances of the two channels are added in parallel, in the case of parallel magnetization the spin-down channel resistance tends to zero for weak diffusion, and a short-circuit effect is created that sends the entire resistance to zero. The magnetoresistance (MR), which is defined as $\Delta R/R_P$, in this case tends to infinity. If instead, we take a more realistic case, with small R_{small} but non zero, we find that,

$$MR = \frac{\Delta R}{R_P} = \frac{R_{AP} - R_P}{R_P} = \frac{(R_{\text{small}} - R_{\text{large}})^2}{4R_{\text{small}}R_{\text{large}}} \qquad (11.10)$$

due to a symmetry situations, $R_{small} = R_{large}$, it is equal to zero. In addition, MR is always positive, which implies $R_{AP} > R_P$. There are two different geometries for the GMR in a multilayer system: the current-in-plane (CIP), where the electric field is applied along the plane of the layer, and the current perpendicular to the plane (CPP) where the electric field is along the normal layer. In both cases, the transport of electrons is substantially different, but it has similar results. The key point of GMR is the asymmetry in the diffusion of electrons due to spin.

11.4 SPIN VALVES

The term spin valve was used to describe an FM/NM/FM system that operates with the principle of GMR. Then, it was used for any device with two FM electrodes that exhibit different resistances for parallel and anti-parallel alignments. The spin valve effect was observed in metal multi-layer, carbon nanotubes, MTJs, graphites, organic and inorganic semicon-ductors. The mechanism for resistance change is not always the same, but it is always originated by the spin of the electron.

To describe a spin valve an optics analogy is use very often. Let us take a beam of light and send it through two polarizers in series, the final intensity will depend on the relative alignment of the two polarizers. If these are parallel, the output light has the highest intensity, while if these are perpendicular, there is no light output. This is due to the fact that the first polarizer only passes vertically polarized light, while the second polarizes only horizontally. By rotating the second polarizer we can obtain a variable intensity. The spin valve is similar, with the spin polarization which plays the role of the polarization of the light, while the ferromagnet is the one of the polarizers.

However, some details do not coincide perfectly. The current does not flow from one ferromagnet to another, but goes parallel to the layers. More-over, there is no spin build up between a ferromagnet and another, while linear polarized light is created between the two polarizers. The analogy works better with "vertical" transport, or systems where the current flows perpendicularly to the layers, what is called CPP, and has been studied and used, in the magnetic tunnel junctions (MTJs).

The spin valve combines two phenomena: the GMR effect and the effect called exchange bias, to obtain commercially exploitable properties. The GMR can take place if there is a relative orientation between two FM layers, whether parallel or antiparallel. We can use the interlayer exchange coupling to obtain, for example, the non-parallel alignment. In addition relying on the IEC to obtain a desired magnetization configuration, a struc-ture with two unpaired FM layers is used, through a relatively thick spacer layer. To get the desired configuration we take advantage of the exchange bias effect: one of the two layers of FM is deposited over an AF layer to obtain a fixed magnetization. The other layer FM is instead free to change its magnetization to get the two different configurations.

Both exchange bias and interlayer exchange coupling are two ways of obtaining and maintaining a magnetization orientation, but they are not the only ones: for example, in what is called the synthetic spin valve, instead of a simple AF layer to lock the magnetization, we use an FM/NM/FM multilayer structure to get the GMR. This results in a stronger coupling and greater thermal stability. Instead, in the so-called pseudo spin valve, it is not based on the AF/FMjunction, but simply on the two FM layers. If the two layers are not coupled, when a magnetic field is applied, the first of the two changes orientation, forming an antiparallel coupling with the second layer, which has not yet changed orientation; in this way, you get a high resistance thanks to the GMR effect.

11.5 MAGNETIC TUNNEL JUNCTION

A very similar device to the valve spins the MTJ, which consists of two ferromagnetic layers, F_1 and F_2, separated by a barrier of an insulating material, also called tunnel barrier. When magnetizations are parallel there is a strong conductance, while for non-parallel magnetizations there is a weak conductance due to spin polarized DOS.

The development of this technology is driven by the ever-increasing demand for hard drives, which exhibits higher ambient temperature resistance. As soon as marketing began, many materials have been developed, the most used being the MgO tunnel barrier with excellent results and performance.

When two conductors are separated by a thin layer of dielectric, electrons can pass from one electrode to the other by a tunneling effect, resulting in electrical conduction. The tunneling phenomenon arises from the electron wave nature, while the nature of the conduction is due to the so-called evanescent state of the electron wave function while it is in the tunnel barrier.

In MTJ there are two ferromagnetic electrodes: within this type of material the current is divided into two conducting channels for spin-up and

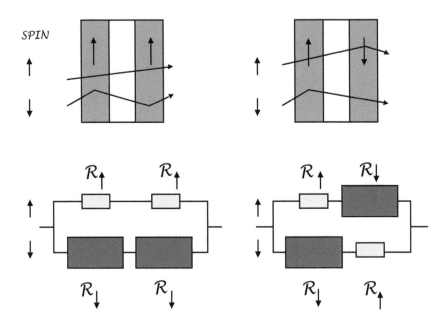

Figure 11.3: A schematic structure of a spin valve, with the two channel CPP resistances model [The electron with spin parallel to the magnetization of the layer has less dispersion than an electron with the opposite spin. In case (b) of non-alignment, both spin channels undergo a high strength resistance and a lower intensity, which in general generates an intermediate intensity resistance. In the case (a) of anti-magnetization, a channel undergoes a great deal of dispersion, the other very small: the short-circuit effect decreases the total resistance.].

spin-electrons. In the tunneling process the spin is retained, so the conductance also depends on the magnetizations of the two electrodes, whether they are parallel or antiparallel. The difference between the two configurations is called tunneling magnetoresistance (TMR). The origin of TMR lies in the difference in spin up and down electron DOS at the Fermi level, denoted E_F. Since they retain spin, the passage occurs only in a band with the same spin orientation. The conductance is proportional to the DOS values of the band with the same spin of the two electrodes. The magnetization switch corresponds to an exchange between the spin bands of one of the two electrodes in the tunneling process, which then causes a difference in conductance because the DOS values have changed.

Let's take for simplicity the case where 100% of spins is polarized at the Fermi level. In parallel configuration, electron spin-up by F_1 due to tunnel effect pass into F_2 spin-up states, this way, high conductance is achieved. In the anti-parallel configuration, electrons with spin-ups cannot pass into F_2 by tunnel effect, because there were no free spin-ups, creating a very low conductance. The TMR was observed by Julliere in 1974 [130], in Fe / Ge oxidized / Co junctions at low temperatures. It was not until the mid-1990s, when TMR was observed in MTJ at room temperature in Al_2O_3 tunnel barriers. More studies led to an increase in TMR from the initial 10%, up to values in the order of 70%.

Unlike GMR, there is no high IEC effect that favors the achievement of anti-parallel alignment. Usually two different FM materials are used, resulting in different coercivities, to achieve non-parallel alignment. Beginning with a negative field, both magnetizations are in a negative direction. When the field grows in module, the magnetization of the material changes, obtaining an alignment. At this point resistance increases with the spin-polarized electron effect of the electrons. When the field reaches higher values, the magnetization of the second material changes, obtaining the parallel alignment again, causing the resistance to decrease.

11.6   JULLIERE MODEL FOR TMR

The two FM layers can be of different materials, so there is no spin-dependent diffusion and the rate of electrons passing from one layer to the other due to the tunnel effect is proportional to the initial and Final DOS at the Fermi level. Let us define D_i and d_i as the density of states at the Fermi level, respectively, of the majority and minority carriers, in the i-th FM layer. For simplicity, $i = 1.2$. Polarization is defined,

$$P_i = \frac{D_i - d_i}{D_i + d_i} \tag{11.11}$$

Taken into account at the Fermi level, in the top layer.

It is defined as $\Delta_i = D_i + d_i$ as the total density of states, from which it is easily obtained,

$$D_i = \frac{\Delta_i}{2}(1 + P_i), \quad d_i = \frac{\Delta_i}{2}(1 - P_i) \tag{11.12}$$

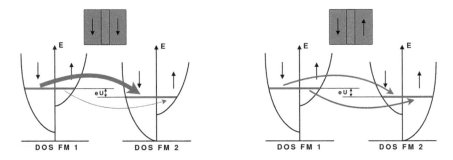

Figure 11.4: A schematic diagram of magnetoresistance tunneling with the Julliere model [In case (a), where the two layers have parallel alignments, it is the majority spin-up population that crossing the tunnel bar finds that the majority position is maintained, and a large number of these electrons can "find place." For minority spins, the situation is similar: they are few in the first layer, and have little room in the second. In the case (b) instead, in the first layer the majority are spin ups, while in the second they are spin down: so there is a large number of spin ups and a spin down limit that try to pass through the tunnel barrier, But on the other side there are many spin downs and a few spin ups. As a result, they succeed in passing a few spin ups for the minority population in the second layer, and a few spin-downs, as a minority population in the first layer. This phenomenon leads to a lot of current in the first case, and therefore low resistance; however, little current and high resistance in the second case.].

Now the TMR is given by,

$$TMR = \frac{\Delta R}{R_P} = \frac{R_{AP} - R_P}{R_P} = \frac{R_{AP}}{R_P} - 1 = \frac{G_P}{G_{AP}} - 1 \qquad (11.13)$$

where R_{AP} is the resistance configuration aligned with the antiparallel magnetization and R_P of the parallel magnetization.

For parallel configuration, the total conductance is the sum of the conductances of the two channels spin, $G_P = G_P^\uparrow + G_P^\downarrow$. On the other hand, the conductance of each channel is proportional to the rate of electrons that, by tunnel effect, pass from one layer to the other, assuming that it is proportional to the initial and final DOS, $G^\uparrow)_P = \alpha D_1 D_2$ and $G_P^\downarrow = \alpha d_1 d_2$, where α is a constant. Similarly, in the case of an antiparallel case, $G_{AP}^\uparrow = \alpha D_1 d_2$ and $G_{AP}^\downarrow = \alpha d_1 D_2$. By putting all together, we obtain,

$$TMR = \frac{G_P}{G_{AP}} - 1 = \frac{G_P^\uparrow + G_P^\downarrow}{G_{AP}^\uparrow + G_{AP}^\downarrow} - 1 = \frac{D_1 D_2 + d_1 d_2}{D_1 d_2 + d_1 D_2} - 1 \quad (11.14)$$

By substituting and using some algebra, we have a clearer expression,

$$TMR = \frac{2P_1 P_2}{1 - P_1 P_2} \qquad (11.15)$$

The Julliere model is a good starting point for analyzing MTJs, but it is a tricky model with strong limitations. First, TMR depends only on the properties of the FM layers and does not take into account the properties of the tunnel barrier. Secondly, there is no distinction between electrons s-band and d-band, which may have different polarization and different effective masses.

11.7 SPIN TORQUE MOMENTUM

The GMR and TMR effects have similar characteristics, such that electron transport properties are strongly influenced by the magnetic configuration of the devices.

At the end of the 1990s, this phenomenon was studied and theorized, assuming a spin torque resulting of spin-dependent reflection or transmission on the FM / NM interfaces, and the retention of angular momentum. The concept of spin torque has earned attention due to its potential for changing magnetization orientation and improve devices such as MRAMs. The switch would be faster than the electrical devices and we can get more integration. The industry always looks for ways of smaller and faster devices, and this can be the answer to both requests.

Spin torque was experimentally observed in two different experiments. In one of them, a system was contacted with a metal through a nano-porous. Measuring the characteristic voltage / current of the device, were observed abrupt changes in the resistance, which correspond to magnetizations. The key role of these experiments is carried out by the very high current density given by nano-contacts. To illustrate the results of the experiments, we use these conventions, the current has a positive sign when an electron runs from the narrowest layer, called free layer, to the broader layer. If the two layers are made up of the same material, the wider layer resists the change of magnetization because strength increases with the volume. When a sufficiently high current is applied, the resistance jumps to a "higher" state. If, however, a sufficiently high negative current is applied, the resistance jumps into "low" state.

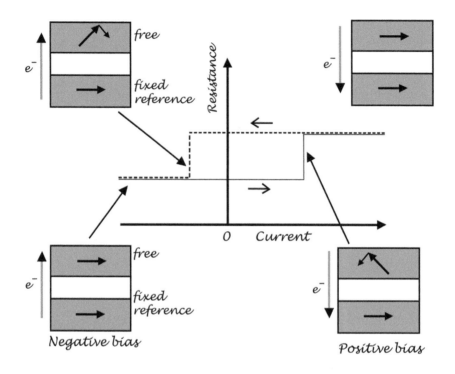

Figure 11.5: The "hysteresis" cycle of resistance due to torque. Positive polarization tends to orient magnetizations in an antiparallel manner while negative polarization is parallel.

From GMR we know that a high resistance corresponds to an antiparallel magnetization, and vice versa. Then a positive current generates an antiparallel alignment while a negative current generates parallel alignment.

11.7.1   Spin Torque Origins

The primary source of the torque is a spin filter effect, whose origins lie in quantum mechanics.

Figure 11.6 shows the geometry of the device. S_1 and S_2 are the spin moments of the magnetic moment of a restricted area in the two ferromagnetic layers F_1 and F_2, the free layer. Since the spin-torque mechanism is based on the retention of the angular momentum S, the magnetization will be treated in terms of angular momentum, which is typically antiparallel.

S_1 is fixed along the $+z'$ axis of our reference system while S_2 lies on the $x'y'$ plane, forming an angle θ with the z' axis. It is convenient to define another system of coordinates (x, y, z) where the z axis is parallel to S_2. In this system the steering vectors will be i, j, k. The spin of free electrons is denoted by s. For simplicity, we assume the transmission and the reflection is complete, the electrons with spin parallel to S_i ($i = 1, 2$) are transmitted completely, electrons with anti-parallel spin are completely reflected. In real systems there is a transmission coefficient that encompasses several effects, including accidental spin polarization and diffusion.

Consider the case where the electrons pass from the reference layer F_1 to the free layer F_2, which we call negative polarization. The spin of the electrons transmitted will be parallel to the direction $+z$. In terms of quantum mechanics $\chi_{in} = |\uparrow\rangle_{z'}$, where the z-axis indicates the quantization axis. When the electron reach the free layer, the transmission and reflection must be calculated using the direction of S_2 as a quantization axis,

$$\chi_{in} = |\uparrow\rangle_{z'} = \cos(\theta/2)|\uparrow\rangle_z + \sin(\theta/2)|\downarrow\rangle_z \qquad (11.16)$$

For the complete transmission and reflection condition, the first term is completely absorbed, while the second reflection, $\chi_T = \cos(\theta/2)|\uparrow\rangle_z$ and $\chi_R = \sin(\theta/2)|\downarrow\rangle_z$.

Introducing the statistical expectation that we call $\tilde{s}$, given by the product of each spin component for the respective direction vector, which is written: $\tilde{s} = \tilde{s}_x i + \tilde{s}_y j + \tilde{s}_z k = \tilde{s}'_x i' + \tilde{s}'_y j' + \tilde{s}'_z k'$. The initial spin is given by,

$$\mathbf{S}_{in} = {}_{z'}\langle\uparrow|\mathbf{S}|\uparrow\rangle_{z'} = \frac{\hbar}{2}\hat{k}' = \frac{\hbar}{2}\sin\theta\,\hat{i} + \frac{\hbar}{2}\cos\theta\,\hat{k} \qquad (11.17)$$

which, divided into the transmitted and reflected components,

$$\mathbf{S}_T = {}_z\langle\uparrow|\cos(\theta/2)\mathbf{S}\cos(\theta/2)|\uparrow\rangle_z = \frac{\hbar}{2}\cos^2(\theta/2)\hat{k} \qquad (11.18)$$

$$\mathbf{S}_R = {}_z\langle\downarrow|\sin(\theta/2)\mathbf{S}\sin(\theta/2)|\downarrow\rangle_z = -\frac{\hbar}{2}\sin^2(\theta/2)\hat{k} \qquad (11.19)$$

The final spin, therefore, as a sum of the transmitted and reflected components, is given by,

$$\mathbf{S}_{fin} = \frac{\hbar}{2}[\cos^2(\theta/2) - \sin^2(\theta/2)]\hat{k} = \frac{\hbar}{2}\cos(\theta/2)\hat{k} \qquad (11.20)$$

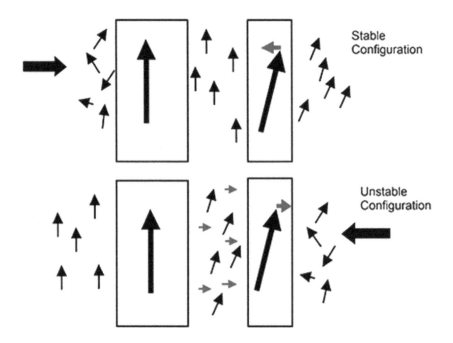

Figure 11.6: The spin–torque mechanism on the free layer when subjected to currents in opposite directions [When the current flows from the fixed layer to the free (negative polarization) layer, spin ups; or fixed layer are used to guide the magnetization of the free parallel layer. If the current flows from the free to the fixed layer instead (positive polarization), the spins are aligned with the free layer: the parallel layer is absorbed by the fixed layer, and reflected that antiparallel, which returns to the free layer. The magnetization switch towards the anti-parallel orientation.].

The change in spin is given,

$$\Delta S = S_{fin} - S_{in} = -\frac{\hbar}{2} \sin \theta \, \hat{\imath} \qquad (11.21)$$

Since the total angular momentum $(s + S_2)$ must be conserved throughout the process, $\Delta s + \Delta S_2 = 0$, which can be rewritten as, $\Delta S_2 = -\Delta s = +\frac{\hbar}{2} \sin \theta i$.

This term represents the change in angular momentum of F_2 of each incident electron, in the case of a perfect spin transfer. A first and partial conclusion is that under negative polarization the torque on S_2 causes a parallel alignment of the free layer magnetization. Additionally, the torque is maximum for $\theta = \pi/2$. When there is an electron current, the torque on F_2 becomes,

$$N_2 = \frac{dS_2}{dt} = \frac{|I|\hbar}{2e}\eta \sin \theta \, \hat{\imath} \qquad (11.22)$$

where I is the current, e is the charge of the electron, η represents the efficiency in spin transport; N_2 in practice is the number of electrons by unit of time multiplied by the angular momentum ΔS_2 per electron.

If we consider the case where an electron runs from the free layer to the reference layer, i.e., the case of polarization, the analysis is very similar, but with a significant difference. With the electrons going to the reference layer, the spin of the reflected electrons will be aligned to the direction $-z'$. Let's take this as the initial spin, $\chi_{in} = |\downarrow\rangle_{z'}$. These reflected electrons will interact with the free layer. At this point, the analysis follows as illustrated previously, but s_{in} brings a negative sign. Then, with positive polarization, the torque on S_2 causes an antiparallel alignment of the free layer magnetization, with that of the reference layer. Of course, in this case the spin torque is written, $N_2 = -\frac{|I|}{e}\frac{\hbar}{2}\eta \sin \theta i$. By combining the positive and negative polarization expressions, we get,

$$N_2 = \frac{I\hbar}{2e}\eta \sin \theta \, \hat{\imath} \qquad (11.23)$$

where the sign of I indicates which polarization is involved. The most important result of the torque analysis is therefore that positive polarization favors antiparallel alignment while a negative polarization favors parallel alignment. This analysis, however, simplified, does not include, for example, the role of magnetization dynamics, although the fundamental results do not change.

11.8 SEMICONDUCTOR SPINTRONICS

Semiconductors are the constituent elements of electronics, thanks to their very versatile transport properties. Only by changing the percentage of doping, the density of carriers can be modified by several orders of magnitude, and it is possible to have carriers both positive (gaps) and negative (free electrons). These properties are used in bipolar junction transistors (BJTs) and field effect transistors (FETs), now used in every field, by amplifying analog signals to logic operations. Some semiconductors have excellent optical-electric properties, used to generate or detect light impulses.

Due to their potentiality, it is of great interest to combine with the spin. In this way, the semiconductor properties are joined to those of the ferromagnets: theoretical models for the realization of spin-FET are introduced, based on spin transport in side channels between source and spin-polarized drain, all controlled by the gate voltage. It can work on hybrid structures with magnetometric and semamagnetic metals, the major problem being the mismatch between the conductivity of the two materials to allow a spin-polarized current to maintain consistency in the junction passage. To count on this, it was proposed to introduce a dielectric barrier, as in some spin-LED models. Another route is represented by the creation of ferromagnetic semiconductors, such as As, with Mn percentages below 10%. These materials have interesting properties, such as being able to control ferromagnetism with a gate voltage, showing a great TMR. However, the critical temperatures of these substances is attuned to $170K$ and requires a greater study.

The study also focuses on creating spin-polarized currents, thanks, for example, to spin-orbit effects such as the spin Hall effect. In these effects, spin-orbital interactions divide the current into the two spin-up and down channels in opposite directions, producing a transverse spin current, all in a diamagnetic conductor. Embedding ferromagnetism in semiconductors with a diluted magnetic doping has demonstrated new magnetic behaviors. In addition, optical studies have demonstrated long spin consistency times, paving the way for solid-state quantum information. The generation of spin polarization in ferromagnetic semiconductors was achieved using spin-injection and reflection.

11.8.1   Ferromagnetic Semiconductors

To join semiconductor properties and ferromagnetism: the diluted magnetic semiconductors (DMS) have been created. In these alloys there is a small percentage of magnetic ions, less than 10%, in a semiconductor that is typically not ferromagnetic. The most commonly used alloys are (In, Mn) As and (Ga, Mn) As. These systems, however, have only ferromagnetic properties below 200 K. Before any technological use, it is necessary to increase Curie's temperature. These materials are ferromagnetic, which is, from an electrical point of view, a hybrid between semiconductors and metals.

The (Ga, Mn) As is grown through MBE at low temperatures (100 − 300°C) with a Mn concentration of between 0 and 10% to obtain a homogeneous alloy. Higher concentrations or temperatures will form MnAs or GaAs clusters that would compromise the operation and properties of the material. In homogeneous alloys, Curie's temperature depends on the concentration of Mn and the growth process, typically it is around 100/150K.

The origin of ferromagnetism in the semiconductors is a topic of interest: as the RKKY pair, the two magnetic moments of the Mn can be coupled indirectly through a charge carrier, in this case a gap, because the manganese is an acceptor. The gap has a mediating role between the magnetic moments of two Mn atoms, explained by the Zener model: when the moments of two manganese atoms interact with the same gap, they align both antiparallel at the gap itself, resulting in parallel alignment. In this way, the lowest level of energy is reached when all moments of manganese are parallel to each other and are antiparallel to gaps.

Although the ferromagnetism mediated by carriers is not yet fully understood, it has been manufactured different devices that use the property. The electromagnetic field control of the ferromagnetism was performed in a FET structure. The ferromagnetism is activated and deactivated with a gate voltage. The effect obtained is to change the concentration of gaps, which in turn increases T_c. In this way it passes from a disordered state to an ordered and vice versa, only by controlling the gate voltage.

These properties have, however, been observed at temperatures far from the ambient temperature, at about 100 K.

11.8.2   Spin Coherence and Spin Orbit Coupling

The primary means by which spin dynamics are measured in the semiconductors are based on a linearly polarized optical beam that is transmitted through a spin population, resulting in the polarization axis rotating at a angle proportional to the spin polarization component along the path of the beam. The measurements are not based on a single pulse, but on a repeated pulse train every 13 ns. If the spin life span is less than 13 ns, independent behaviors are monitored, but if the impulses are higher, they interfere with each other and more complex phenomena are observed. Several experiments have been conducted on GaAs that have produced unexpected results. With a doping of $1016cm^{-3}$, the lifespan of the spin found is about 100 ns at 5 K. The doping addition is not trivial, since in a non-doped system the lifespan is limited by recombination time, and once the electrons are recombined with the gaps, they are no longer carriers.

In doped systems, free electrons (or gaps) are present at equilibrium, so spin polarization can persist even after recombination. Spin-orbit phenomena come into play at doping levels, which heavily restrict spin-time coherence. The role of spin-orbit coupling is extremely important in semiconductors, and provides both desired and undesired effects. On the one hand, it limits the time of spin coherence, so it is preferable to have a low spin-orbit coupling. On the other hand it is indispensable to generate, reveal, and manipulate optically the spin.

This is a relativistic effect that arises when we consider the same situation from two different points of view, which we call "laboratory" and "electron." In the first point we consider an electron traveling near a resting nucleus positively charged. The core creates an electric field, but not a magnetic field, since it is at rest. From the second point of view, the core moves. His motion produces a current that in turn produces a magnetic field, due to the law of Biot-Savart. This field, which we call H_{SO}, interacts with the magnetic moment (m) of the spin trough the Zeeman's energy term: $E_{Zeeman} = -m \cdot H_{SO}$. From its relativistic origin, the H_{SO} field increases with the velocity of the electron. Then, in the laboratory system there is an electric field $\vec{E}$, while in the electron system an internal magnetic field H_{SO} is generated. The shape of this field is known as a Rashba spin-orbit coupling and is written as $\vec{H}_{SO} \approx \vec{E}\vec{k}$, where $\vec{k}$ is the electron wave vector.

Given a population of spin-polarized electrons, there is a momentum distribution and hence a distribution of H_{SO}. Each electron stars a precession along a different axis, causing the net spin polarization to be canceled. This momentum-shifting mechanism is named after Elliot-Yafet. As mentioned above, spin life spends with increasing doping, due to spin-orbit coupling effect. The conduction band is filled, and the electron wave carriers increase. This causes a higher coupling, which results in a short-ening of the coherence time. Although the dephasing effect is strongly undesirable, the spin-orbit coupling can be used to manipulate the spin. If a spin population is moving at a certain average drift speed, there is an average non-zero waveform vector $\langle \vec{k} \rangle$. Thanks to the Rashba Coupling, this creates a not zero internal field $H_{SO} \approx E \times \langle \vec{k} \rangle$ in the presence of an electric field. The Zeeman splitting and spin precession resulting from this field is named after Rashba. From these studies, Datta and Das proposed a spin-transistor model based precisely on this effect in 1990. This device consists of a drain and a ferromagnetic source that injects and receive spin in a two-dimensional electron gas channel (2DEG). The idea resembles the spin valve, but with a fundamental difference, the presence of an electro-static gate that can generate an electric field perpendicular to the 2DEG. Since the electrons move from the source to the drain, the Rashba effect generates an average internal $\langle H_{SO} \rangle$ field in the direction perpendicular to the electric field. Since even the spin is perpendicular to $\langle H_{SO} \rangle$, it will precess with a frequency that depends on the electric field, controlled by the gate voltage.

$$\omega_L = \frac{g\,\mu_B |\langle \mathbf{H}_{SO} \rangle|}{\hbar} \sim \frac{g\,\mu_B |\langle \mathbf{k} \rangle||\mathbf{E}|}{\hbar} \tag{11.24}$$

When the spin reaches the drain, the relative orientation between drain magnetization and spin determines the source-drain current: maximum current for parallel alignment, minimum for non-parallel alignment. The final orientation depends on the precession frequency, which depends on the gate voltage. The current source-drain/gate voltage characteristic assumes a zigzag shape.

It should be noted that a small gate voltage change can lead to major changes in the source-drain current. The proposal of the Datta-Das spin-transistor has given strength to research in the field of spintronics in semi-conductors, although it must still be realized, it took many years to show that the fundamental components of the project were achievable experi-mentally.

11.8.3 Spin-LED

A big challenge of spintronics is the injection of spin-polarized electrons by a ferromagnet into a diamagnetic semiconductor. While studies of GMR with the perpendicular plane current (CPP) have shown spin injection from a ferromagnet to a diamagnetic metal, a demonstration of the same effect in a semiconductor is more complicated. The major difficulty is the lack of correspondence between ferromagnetic conductivity and semiconductor conductivity. Spin-dependent LEDs (spin-LEDs) provided the first definitive spin injection demonstration in a semiconductor. Instead of a ferromagnetic metal as a spin injector, ferromagnetic semiconductors such as (Ga, Mn) As, or paramagnetic semiconductors such as (Be, Mn) ZnSe were used. In experiments with (Ga, Mn) As a voltage is applied to inject holes from the (Ga, Mn) As (which is of type p) in the intrinsic GaAs. These recombine in the InGaAs with non-polarized electrons from a type n GaAs injector.

The lightness of light emitted depends on the spin polarization of the injected holes. By measuring the circular polarization of the light as a function of the magnetic field and the temperature, it is found that the emitted light corresponds to the magnetization of the layer (Ga, Mn) As. This demonstrates the spin injection of the ferromagnetic layer (Ga, Mn) As in the GaAs diamagnetic layer. By recombining spin-polarized holes with non-polarized electrons, circular polarized light emission results. This polarization depends proportionally on the direction of observation, this fact is used as an indicator of the injected spin polarization. From the demonstration of spin injection in this semiconductor structure, Schmidt provided a model that explains the success of these experiments. The spin polarization is assumed as

$$\mathbf{S} = \beta \frac{\sigma_N}{\sigma_F} \frac{\lambda_F}{\lambda_N} \tag{11.25}$$

where β is the spin polarization of the ferromagnet, σ_F and σ_N are the conductivity of the ferromagnet and non-magnet respectively, λ_F and λ_N are the wavelength of the spin diffusion of the ferromagnet and non-magnet, respectively. The problem found of injecting from a ferromagnetic metal to a non-magnetic semiconductor is found in the ratio σ_N / σ_F which is in the order of 0.001. Although a ferromagnet may have $\beta \approx 30\%$, this would generate a polarization of about 0.03 % in the semiconductor. Applying a relatively large magnetic field, the Zeeman effect creates a total polarization of carriers in the DMS layers. If the thickness of this layer is sufficiently small, less than the length of diffusion in the material, the spins entering in the transport layer are completely polarized. Rashba has also offered a solution to the lack of matching between conductors in the metal/ semiconductor interface, introducing a tunnel barrier between the two, having a greater resistance to that of the semiconductor. In some metal / semiconductor systems there is a potential barrier from the semiconductor side, known as the Schottky barrier. In the Fe/GaAs interface, the presence of the Schottky barrier makes possible to inject spin from Fe to GaAs without the introduction of a tunnel barrier. Additionally, using doped gradients to modify the Schottky barrier we can achieve different performance in spin injection.

CHAPTER 12

QUANTUM DOTS

The quantum dots (QDots), also known as low-dimensionality heterostructures or artificial atoms, are solid semiconductor structures, composed mainly of Gallium Arsenide (GaAs), Arsenide de Galium-Aluminum (GaAsAl), Cadmium Selenide (CdSe), or Galena (PbS). The QDots act as a box that confines particles, whether electrons, holes or excitons, the numbers can be controlled by applying a potential across two metal electrodes connected to the system. The confinement of the particles creates a discrete quantization of energy levels, producing changes in the electrical and optical properties of the system, permitting new possibilities in the design of artificial atoms and molecules.

In normal semiconductor devices it is necessary to use multiple electrons to define a certain electron state. In the quantum points, its electron state is determined by a small number of carriers.

The discovery of these heterostructures dates back to the early eighties. Due to the heating of the samples, some small particles of semiconductors precipitated into the glass, causing anomalous optical behavior due to the quantum confinement of the electrons in these crystals.

One way to understand this reasoning is to imagine an electron trapped in a box. This system is modeled as a well of infinite potential, of width l_i, where the energy levels are of the form

$$E_i = \frac{\hbar^2 n^2 \pi^2}{2\mu l_i^2} \, , \quad n = 1, 2, 3, \ldots \, ; \quad i = 1, 2 \tag{12.1}$$

If the size of the confining box of the electron decreases, the lower energy level of the electron increases. In the case of semiconductor nanocrystals, this level corresponds to the threshold energy of optical absorption.

12.1   STRUCTURE OF QUANTUM DOTS

As mentioned before, Qdots have dimensions, and the number of atoms between the molecular atomic level and bulk materials with a band interval depends in a complex way upon a number of factors, including the type of bond and strength with the nearest neighbors. For isolated atoms, acute and narrow luminescent emission peaks are observed. However, a nanoparticle, composed of about $100\text{-}10,000$ atoms, presents distinct optical spectra narrow line. This is why, QDots atoms are often described as artificial atoms. Current research aims using the unique optical properties of quantum dots in devices, such as light emitting devices (LEDs), solar cells and biological markers. The most fascinating change of Qdots with a particle size < 30 nm are the drastic differences in optical absorption, exciton energies and recombination of electron-hole pairs. The size dependence comes from changes in surface-to-volume ratio with size and quantum confinement effects. However, Qdots present different colors with change in size. Figure 12.1 shows the change in the photoluminescence (PL) emission in color with the size of CdSe Qdots.

12.1.1   Size in Relation to the Density of States

A very unique property of the Qdots is the quantum confinement, which modifies the DOS near the edges of the band. The quantum confinement effects are observed when the size is sufficiently small that the spacing of the energy level of a nanocrystal exceeds kT, where k is the Boltzmann constant and T is the temperature. The energy differences $> kT$ limit the mobility of electrons and holes in the crystal. Of the many properties that are dependent on the size of Qdots, two are of particular importance. The first is an increase in band energy when the diameters of the nanoparticles are less than a particular value, which depends on the type of semiconductor. This is called a quantum confinement effect. This effect adjusts the energy gap with changes in the Qdot size. The energy of the band space also depends on the composition of the semiconductors as well as

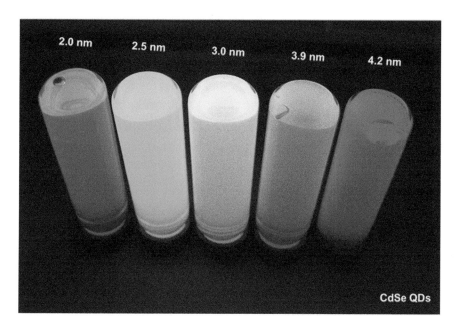

Figure 12.1: Emission colors from small (blue) to large (red) CdSe Qdots excited by a near-ultraviolet lamp.

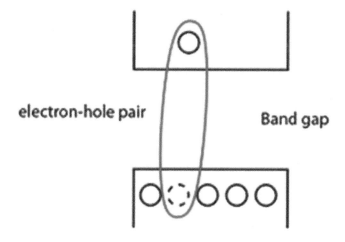

Figure 12.2: Electron-hole pair induced by a electromagnetic radiation or a external potential.

the size. The second important property is the observation of discrete and well-separated states of energy due to the small number of atoms in Qdots with respect to the bulk. This leads to the electronic states of each energy level having wave functions that are more atomic-like. Since the Qdots solutions for the Schrödinger wave equation are very similar to those of the nucleus-bound electrons, Qdots are called artificial atoms and acute atom-like emission peaks are possible. Typical intraband bandwidth spaces for Qdots are between 10 and 100 MeV. The band gap can also be granted by combining the core of the Qdots.

12.1.2 Phases and Phase Transitions

The II–VI compound semiconductors include zinc, cadmium and/or mercury combined with anionic oxygen, sulfur, selenium and tellurium. These semiconductors generally crystallize in hexagonal crystal structures both face centered. For example, ZnO and ZnS equilibrium crystal structures are hexagonal, although ZnS often has a cubic or metastable cubic or hexagonal cubic structure. The II–VI compound semiconductors may exhibit very good luminescence because they have a direct band gap. In addition, many II–VI semiconductors are often Used as a host for the luminescent activators, for example, ZnS doped with Mn^{2+} which emits yellow Light. An emission close to the exciton band can be observed from II–VI semiconductors, particularly at low temperature from those materials with low exciton binding energy.

Qdots exhibits a solid-solid phase transition as bulk semiconductors. Phase transitions in bulk materials can be induced by variable pressure, temperature and composition. The bulk CdSe may have either a hexagonal or a cubic structure with a direct or indirect band deviation, respectively. Above a pressure of $\approx 3 GPa$, the bulk semiconductor CdSe can be reversibly converted from low pressure hexagonal to high pressure cubic structures [113]. The low intensity optical emission of the cubic form of CdSe is in the near infrared spectral region (NIR) at 0.67 eV (1.8 μm).

Structural transformation has also occurred in CdSe Qdots [111], [112]. The ratio of the oscillator strength between the direct and indirect structures did not change with the size of Qdot.

12.1.3 Doping in Quantum Dots

Doping in Qdots is an important aspect when used in several technological applications, particularly optoelectronic, magnetic, biological and spintronic applications. These impurities, called activators, disrupt the structures of the band by creating local quantum states that lie within the band gaps.

In the Qdots, the dopants prove to be auto-ionized without thermal activation due to the quantum confinement.

When quantum confinement energy exceeds the Coulomb interaction between the carrier holes or electrons, and an impurity, the auto-ionization occurs. Several transition elements such as Cr, Mn, Fe, Co, Cu and Ag have been doped in Qdots for various applications. The optical properties of Qdots can vary by modifying the quantities and the positions of the dopants. In the optoelectronic application of Qdots, doping can play an important role. Conduction in doped Qdot films depends on the uniformity of the size of Qdots and the proximity of neighboring Qdots, so that the orbital overlap between Qdots is maximized. The optical properties of Qdots can be improved by doping in Qdots.

12.1.4 Quantum Dot Alloying

By keeping the size constant of a Qdot, the band gap can be designed by the alloy of the core. The composition of the base materials or the ratio of the alloying materials can modify the optoelectronic properties of a Qdot. Such a process has been thoroughly investigated over the years. The process is particularly interesting because: these semiconductor nanostructures provide different and nonlinear optoelectronic properties and Qdot alloyed with multiple semiconductors have mixed or intermediate optoelectronic properties. The efficiencies of PL emissions can be improved by minimizing bulk and surface defects, and considering a narrow half-maximum width (FWHM) of PL. For example, a weak quantum confinement regime, $Zn_xCd_{1-x}S$ Qdots were synthesized using a wide band interval and small Bohr radii ZnS and CdS semiconductors. Since these Qdots are in the weak quantum confinement regime, the non-homogeneous widening of PL due to size fluctuation is considerably reduced. These Qdots are also used in many applications, including biological imaging [116].

12.1.5 Quantum Confinement Effects and Prohibited Bandwidth

Quantum confinement generally results in a widening of the bandwidth with a decrease in the size of the Qdots. The forbidden bandwidth in a material is the energy required to create an electron and a hole at rest, i.e., with zero kinetic energy, at a distance sufficiently separated. That is, its Coulomb interaction can be neglected. If one approaches the other carrier, they may form an electron-hole bonded pair, that is, an exciton, whose energy is a few meV less than the forbidden bandwidth. This exciton behaves like a hydrogen atom, except that a hole, and not a proton, forms the nucleus. Obviously, the mass of a hole is much smaller than that of a proton, which affects the solutions of the Schrödinger equation. The distance between the electron and the hole is the Bohr exciton. If m_i and m_h are the effective masses of electrons and holes, respectively, the Bohr radius of the exciton of bulk semiconductors can be expressed by (12.2), where ϵ, $\hbar$, and e are the optical dielectric constant, Planck's constant reduced and electron charge, respectively.

$$r_B = \frac{\hbar^2 \varepsilon}{e^2} \left(\frac{1}{m_e} + \frac{1}{m_h} \right) \tag{12.2}$$

If the radius (R) of a Qdot approaches to the Borh radius r_B, that is, $R \sim r_B$, or $R < r_B$, the motion of the electrons and the holes are spatially confined to the dimension of the Qdot, which causes an increase in the energy of the exciton transition and the observed blue shift in the Qdot band gap. The Bohr radius of the exciton is a threshold value, and the confining effect becomes important when the radius of the Qdot is smaller. For the small Qdots, the energy of binding of excitons and biexciton binding energy (exciton-exciton energy) is much greater than for bulk materials. Note that for a material with a relatively higher or lower ϵ and m_h, r_B is higher. Two detailed theoretical approaches are used to better predict the properties of excitons, specifically the effective mass approximation model (EMA) and the linear combination of atomic orbit theory (LCAO).

12.1.5.1 *Effective Mass Approach Model*

This approach, based on the model of a particle in a box, is the most widely used model to predict quantum confinement. It was first proposed by Efros in 1982 [119] and subsequently modified by Brus in 1986 [120]. It assumes a particle in a potential well with an infinite potential barrier at the particle boundary. For a free particle to assume any position in the box, the relation between its energy (E) and the wave vector (k) is given by

$$E = \frac{\hbar^2 k^2}{2m^*} \tag{12.3}$$

In the EMA model, it is assumed that this ratio is maintained for an electron or hole in the semiconductor, therefore the energy band is parabolic near the edge of the band. The displacement of the band energy (ΔEg) due to the confinement of the exciton in a Qdot with a diameter R can be expressed as,

$$\Delta E_g = \frac{\hbar^2 \pi^2}{2\mu R^2} - \frac{1.8e^2}{\varepsilon E} = \frac{\hbar^2 \pi^2}{2R^2}\left(\frac{1}{m_e} + \frac{1}{m_h}\right) - \frac{1.78e^2}{\varepsilon E} - 0.248 E_{Ry}^* \tag{12.4}$$

where μ is the reduced mass of an electron-hole pair and $E_{R_y}^*$ is the Rydberg energy.

The first term of (12.4) represents a relation between quantum-location "particle-in-a-box" energy or the confining energy and radius of Qdot (R), while the second term shows the interaction Columbic energy with an $1/R$ dependency. The Rydberg energy term is independent of size and is generally neglected, except for semiconductors with small dielectric constant. Based on (12.4), the first excitonic transition increases as the Qdot (R) radius decreases and shifts the Columbic excited electronic state to lower value. However, the EMA model is not valid into the small Qdot regime, because the $E - k$ relationship can no longer be approximated as parabolic.

12.1.5.2 *Linear Combination of Atomic Orbital Theory-Molecular Orbital Theory*

A model based on a linear combination of orbital orbital-molecular orbital (LCAO-MO) provides a detailed basis for predicting the evolution of the electron structure of clusters of atoms or molecules to Qdots to bulk materials, and predicting the dependence of the gap in the size of the crystals.

In Si diatomic molecule, the atomic orbitals (AO) of two individual atoms combine to produce molecular orbital bonding and anti-bonding. In this approach, nanometric Qdots are considered as large molecules. As the number of atoms increases, the discrete energy band structure changes from large steps to small energy steps, that is, to a more continuous energy band. The occupied molecular orbital quantum states, equivalent to the valence band, and are called the highest occupied molecular orbital levels (HOMO). Unoccupied anti-bonding orbitals, equivalent to the conduction band, and are called the lowest unoccupied molecular orbital levels (LUMO). The energy difference between the top of the HOMO and the bottom of the LUMO increases and the bands are divided into discrete reduced energy levels mixing of AOs for a small number of atoms. Therefore, the small size of the Qdots results in quantized electronic band structures intermediate between the atomic/molecular and bulk crystalline MOs.

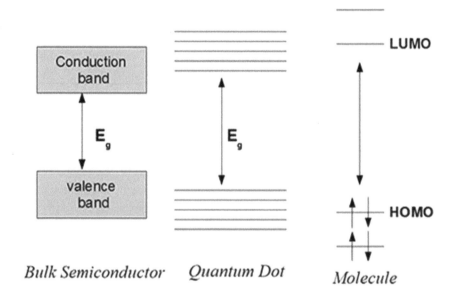

Figure 12.3: The occupied molecular orbital quantum states are equivalent to the valence band (HOMO), and the unoccupied anti-bonding orbitals are equivalent to the conduction band (LUMO).

Compared to the effective mass approximation, the LCAO-MO model provides a methodology to calculate the electronic structure of much smaller Qdots. In contrast, this method cannot be used to calculate the energy levels of large Qdots because of the mathematical complexity and limitations of computer systems. However, the degree of quantum confinement is determined by the ratio of the radius of a Qdot to bulk excitonic Bohr radius. In crystals of larger size than the excitonic Bohr diameter, the semiconductor crystals show a confluence of translational motion of the fully coupled exciton due to a strong Coulombic interaction between the electron and the holes, i.e., they exhibit a confining behavior of single particle regime. In the intermediate size range ($R \leq r_B$), the transition energies of the photoexcited carriers in the crystal are determined by the relative forces of the confining kinetic energy and the electron-hole interaction.

The band gap of the Qdots can be determined by electrochemical measurement using Qdots films. Cyclic voltammetry is often used to determine the oxidation and reduction potential of the Qdots film to be measured using a standard three-electrode cell.

At this point, we can ask about the proper size, so the nanocrystal need to have for this phenomenon to be visible. In the vacuum, the confinement effects are noticeable when the electron is trapped in a volume with a radius of 10 Å. Likewise, due to the confinement, the electrostatic potentials of the crystalline lattice atoms that constitute the system overlap, producing a medium where the electron waves propagate with less resistance than in the free space. In this medium, the effective mass of the electron is smaller than the actual mass. For semiconductor materials such as Gallium Arsenide (GaAs), the effective mass is around 7% of what would be in the vacuum and for Silicon (Si) is 14%, so if it is created a quantum dot with these materials, when measuring the displacement of the energy of absorption, the quantum dot of GaAs presents a greater displacement than a quantum dot of Si.

12.1.6   Theoretical Characteristics of the Physical System

Let us focus on a quantum dot of GaAs/AlGaAs of radius ρ, which confines in its interior an electron. The whole system is externally affected by a homogeneous and constant magnetic field of the form $\vec{B} = B\hat{z}$ in the direction parallel to the z-axis. The electron confinement potential in the z-axis is considered to be an infinite well, whereas on the xy-plane it is taken as a two-dimensional isotropic harmonic potential. The electrostatic interactions of the different GaAs atomic nuclei on the electron are modeled under the effective mass approximation, assuming a description at temperature $T = 0K$, in order to consider only the states of low energy. The parameters used refer to quantum dots made with the electron two-dimensional gas technique (2DEG).

The Hamiltonian that describes this system, including the interaction due to the external magnetic field and the electron spin is,

$$ H = \frac{1}{2\mu} \left[p + \frac{q_e}{c} A \right]^2 + \frac{\mu\omega_0^2}{2} \left(x^2 + y^2 \right) + u(z) - \frac{1}{2} g_s \mu \cdot \sigma \mathbf{B} \quad (12.5) $$

where $\vec{A}(r) = (B/2)(-y\hat{x} + x\hat{y})$ is a vector potential from which it is possible to derive the magnetic field $\vec{B}$ and satisfies the condition of $\nabla \cdot \vec{A}(r) = 0$, $\hat{P}$ is the total momentum of the system, $(1/2\mu)|\hat{P}+(e/c)\vec{A}(r)|^2$ is the kinetic energy, including the associated magnetic field interaction. μ is the effective mass of the electromagnetic oscillator on the GaAs, the term $\mu\omega_0^2/2(x^2 + y^2)$ is the confinement potential in the direction of the z axis which is modeled by an infinite-well potential and the last term represented is the Zeeman coupling, where g_s is the gyromagnetic factor, $\vec{\sigma}$ are Pauli's matrices. The methodology used to determine the energy eigenvalues end eigenfunctions associated with H consists of separating (12.5) into two independent Hamiltonians. The first one, denoted by H_0, which does not include the interaction with the electron's spin and is composed of the first three terms of (12.5), whereas the second contribution H_s we consider the effect of spin and is the fourth term of Eq. (12.2).

12.1.7 Quantum Dot Without Spin Effects

The Hamiltonian that represents the quantum point without the effects of spin is

$$ H_0 = \frac{1}{2\mu} \left[p + \frac{q_e}{c} \mathbf{A}(r) \right]^2 + \frac{\mu\omega_0^2}{2} \left(x^2 + y^2 \right) + u(z) \quad (12.6) $$

Considering that $[x, p_y] = [y, p_x] = 0$, representing the cyclotron frequency as $c = eB/c\mu$, the projection of the orbital angular momentum on the z-axis as $L_z = xp_y - yp_x$ and $^2 = \frac{2}{0} + (1/4)_c^2$, the third term takes the form,

$$H_0 = \frac{p_x^2}{2\mu} + \frac{p_y^2}{2\mu} + \frac{p_z^2}{2\mu} + \frac{1}{2}\omega_c L_z + \frac{1}{2}\mu\Omega^2\left(x^2 + y^2\right) + u(z) \quad (12.7)$$

The Eq. (12.6) separates into two independent Hamiltonian variables. The first term describes the behavior of the electron in the xy plane, while the second represents a potential well confinement of infinite walls in the z-axis. Therefore,

$$H_0 = H_\perp + H_\parallel \quad (12.8)$$

where

$$H_\perp = \frac{p_x^2}{2\mu} + \frac{p_y^2}{2\mu} + \frac{1}{2}\omega_c L_z + \frac{1}{2}\mu\Omega^2\left(x^2 + y^2\right) \quad (12.9)$$

and

$$H_\parallel = \frac{p_z^2}{2\mu} + u(z) \quad (12.10)$$

p_z is the momentum in the direction z and $u(z)$ the confinement potential in z.

12.1.7.1 Hamiltonian in the xy-Plane

The Hamiltonian of (12.9) represents a two-dimensional harmonic oscillator with an external magnetic field interaction. The process for determining the values of $H_\perp$ is analogous to the solution of the one-dimensional quantum harmonic oscillator. In this case it is necessary to construct two couples of operators of creation–destruction, because it is a two-dimensional system. Defining $\beta = \sqrt{\mu\Omega/\hbar}$, the first set of operators is expressed as a function of position and momentum, thus,

$$a^\dagger = \frac{1}{2}\left(\beta x - \frac{ip_x}{\beta\hbar}\right), \quad b^\dagger = \frac{1}{2}\left(\beta y - \frac{ip_y}{\beta\hbar}\right) \quad (12.11)$$

$$a = \frac{1}{2}\left(\beta x + \frac{ip_x}{\beta\hbar}\right), \quad b = \frac{1}{2}\left(\beta y + \frac{ip_y}{\beta\hbar}\right) \quad (12.12)$$

whereby the Hamiltonian $H_\perp$ takes the form,

$$H_\perp = \hbar\omega\left(a^\dagger a + b^\dagger b + 1\right) + \frac{i\omega_c\hbar}{2}\left(ab^\dagger - a^\dagger b\right) \quad (12.13)$$

Introducing a second set of annihilation and creation operators,

$$A^\dagger = \frac{1}{\sqrt{2}}\left(a^\dagger + ib^\dagger\right), \;\; B^\dagger = \frac{1}{\sqrt{2}}\left(a^\dagger - ib^\dagger\right) \tag{12.14}$$

$$A = \frac{1}{\sqrt{2}}\left(a - ib\right), \;\; B = \frac{1}{\sqrt{2}}\left(a + ib\right) \tag{12.15}$$

reduce to,

$$H_\perp = \hbar\Omega\left(A^\dagger A + B^\dagger B + 1\right) + \frac{\omega_c \hbar}{2}\left(A^\dagger A - B^\dagger B\right) \tag{12.16}$$

Denoting the product of operators $A^\dagger A$ and $B^\dagger B$ as N_A and N_B with eigenvalues n_A and n_B respectively, this Hamiltonian becomes,

$$H_\perp = \hbar\Omega\left(N_A + N_B + 1\right) + \frac{\omega_c \hbar}{2}\left(N_A - N_B\right) \tag{12.17}$$

If $\omega_A = \Omega + (1/2)\omega_c$ and $\omega_B = \Omega - (1/2)\omega_c$, when the Hamiltonian is applied to an arbitrary state of the system, the energy eigenvalue is expressed as,

$$E_\perp = \hbar\omega_A\left(n_A + \frac{1}{2}\right) + \hbar\omega_B\left(n_B + \frac{1}{2}\right) \tag{12.18}$$

The energy of the quantum point in the xy plane depends on the eigenvalues of the operators N_A and N_B, as well as the frequencies ω_A and ω_B, which are implicitly associated with the confinement frequency of the dot and the external magnetic field applied on the system. Defining $n = n_A + n_B$ and $m = n_A - n_B$, Eq. 14 takes the form,

$$E_\perp = \hbar\Omega(n + 1) + \frac{1}{2}\omega_c \hbar m \tag{12.19}$$

Being the energy in the xy plane associated with an electron confined in a quantum dot.

To determine the wave function, it is necessary to make use of the properties presented by the operators N_A and N_B, which are exactly the same as those presented by the operator N for a one-dimensional oscillator. Any state of the system is described as a fundamental state function, on which such operators act. This relation is described by,

$$|\varphi_{n_A, n_B}\rangle = \frac{1}{\sqrt{n_A! \, n_B!}}(A^\dagger)^{n_A}(B^\dagger)^{n_B}|\varphi_{00}\rangle \tag{12.20}$$

To determine the form of $|\phi_{00}\rangle$, we replace the operators A, $A^\dagger$, B, $B^\dagger$ by the operators a, $a^\dagger$, b, $b^\dagger$, additionally, making a change of variable $z = X + iy$ and $z^* = x - iy$ and taking as reference the fact that $A|\phi_{00}\rangle = 0$, $B|\phi_{00}\rangle = 0$ it is obtained,

$$\left(\frac{\beta}{2}z^* + \frac{1}{\beta}\partial_z\right)\psi_{00}(z, z^*) = 0 \tag{12.21}$$

$$\left(\frac{\beta}{2}z + \frac{1}{\beta}\partial_{z^*}\right)\psi_{00}(z, z^*) = 0 \tag{12.22}$$

Its solution being the wave function in the ground state

$$|\psi_{00}\rangle = \sqrt{\frac{\beta^2}{\pi}}e^{-\frac{\beta^2}{2}(x^2 + y^2)}. \tag{12.23}$$

The (12.10), with a potential $u(z)$ of the form

$$u(z) = \begin{cases} \infty, & z < -\frac{1}{2} \quad \text{or} \quad z > \frac{1}{2} \\ 0, & -\frac{1}{2} < z < \frac{1}{2} \end{cases} \tag{12.24}$$

Describes the problem of an infinite potential well. The system eigenstate is given by,

$$|\varphi_{n_z}\rangle = \begin{cases} \sqrt{\frac{2}{L}}\cos(k_n z), & k_{n_z} = \frac{n_z\pi}{L} \quad \text{and} \quad n_z = 1, 3, 5, \ldots \\ \sqrt{\frac{2}{L}}\sin(k_n z), & k_{n_z} = \frac{n_z\pi}{L} \quad \text{and} \quad n_z = 2, 4, 6, \end{cases} \tag{12.25}$$

and the eigenvalues are

$$E_\parallel = \frac{n_z^2\pi^2\hbar^2}{2\mu L^2}. \tag{12.26}$$

12.1.7.2 *Quantum Dots with Spin Effects*

To consider spin effects in this system, it is necessary add to the Hamiltonian H a term representing the interaction between the electron's spin S with a homogeneous magnetic field in the z direction, not including the spin-orbit interaction. This term is represented by,

$$H_s = -\mu_s \cdot \mathbf{B} \tag{12.27}$$

where $\mu_s = -(g_s \mu_B S/\hbar)$, $\mu_B = (e\hbar/2\mu)$ and $S = (\hbar/2)\sigma$. Replacing in the above equation, we get,

$$H_s = \frac{g_s q_e \hbar}{4\mu} \sigma \cdot \mathbf{B} \tag{12.28}$$

where $|\phi_{m_z}\rangle = |\pm\rangle$, in such a way that,

$$\sigma_z |\varphi_{m_z}\rangle = \sigma|\pm\rangle = \pm|\pm\rangle = \pm|\varphi_{m_z}\rangle \tag{12.29}$$

Because the magnetic field is in the z-direction. Taking into account the angular momentum of the spin alone ($g_s = 2$), the eigenvalue equation associated with this system is,

$$H_s |\varphi_{m_z}\rangle = \frac{q_e \hbar}{2\mu} \sigma_z \mathbf{B}|\pm\rangle \tag{12.30}$$

When acting on a state $|\phi_{m_z}\rangle$ gives an eigenvalue of energy given by,

$$E_s = \pm \frac{q_e \hbar}{2\mu} B \tag{12.31}$$

The state vector for the electron located at the quantum dot, under a magnetic field interaction in the z direction, can be expressed as the tensor product between the state vector corresponding to the two-dimensional harmonic oscillator, the eigenvector of the infinite potential well and the electron state vector with spin. Therefore, the total vector state is,

$$|\psi\rangle = |\varphi_{n_A, n_B}\rangle \otimes |\varphi_{n_z}\rangle \otimes |\varphi_{m_z}\rangle = |\varphi_{n_A, n_B, n_z, m_z}\rangle \tag{12.32}$$

Likewise, the total energy self-value of the system is,

$$E = \hbar\Omega(n+1) + \frac{\hbar\omega_c}{2}(m) + \frac{n_z^2 \pi^2 \hbar^2}{2L^2 \mu} \pm \frac{q_e \hbar}{2\mu} B \tag{12.33}$$

CHAPTER 13

MAGNETIC RESONANCE

The origin of the spin concept was probably the most complicated of all the quantum physics. Indeed, after Bohr's triumph in 1913, Bohr and Sommerfeld's "old quantum theory," based on the idea of quantum restrictions applied to classical magnitudes, gave a unified explanation of the spectroscopic data [131]. Nevertheless, there was an enormous accumulation of inexplicable facts, even of paradoxes. Abnormal Zeeman effect, spectral line cleavage, electronic layers of complex atoms, Stern and Gerlach's experiments, and so on. For the first time there appeared a purely quantum quantity, without any classical analogy. The spin $1/2$ conditions the whole physical world.

Atomic physics shows that it is not possible to account for the observed effects if we assume that the electron, a point particle, and the three degrees of freedom of translation in space that we have considered up to now. A quantity based on experimental observations and theoretical arguments admits the existence of an internal degree of freedom: its own kinematic momentum. It is a magnitude without a classical analogue: any modeling of this own kinetic momentum in the form of a rigid rotator is impossible. In other words, the electron, a punctual particle, "spins" on itself. This "rotation" is purely quantitative.

In quantum relativistic mechanics, the structure of the Lorentz group reveals the kinetic momentum to any particle as an intrinsic attribute, which is defined in the same way as its electric charge and its mass. We are interested here in the case of the spin $1/2$, that is to say a proper kinetic momentum corresponding to eigenvalues $j = 1/2$, $m = \pm 1/2$. Generally speaking, the spin of a particle is called proper or intrinsic kinematic momentum as opposed to its orbital kinematic momentum. The spin can then take the whole range of the values associated with orbital momentum, $j = 0$ for the meson π, $j = 1$ for the photon or the deuteron, $j = 3/2$ for some elementary particles in the nuclei, etc. Spin $1/2$ is a truly quantum quantity, and the representation that each one makes for this concept is a personal affair, using arrows, pointing fingers, for example. We shall then return to the relation between momentum and magnetic moment, in order to arrive at a phenomenon of great practical importance, the magnetic resonance.

13.1   SPIN $\frac{1}{2}$ HILBERT SPACE

The degree of freedom associated with the proper kinetic momentum of a particle is manifests itself explicitly by physical quantities that are the projections on three axes x, y, z of this momentum, and also as any function of these three magnitudes. The fundamental property of a spin $1/2$ particle is that when it is projected, the henceforth called spin, along any axis, the only one values observed are $+\hbar/2$ and $-\hbar/2$. From this experimental result, comes the fact that by measuring the square of any component of the spin, we find only one value $\hbar^2/4$, with probability one. Consequently, the measurement of the square of the spin $S^2 = S_x^2 + S_y^2 + S_z^2$, gives the result $S^2 = 3\hbar^2/4$, whatever the spin state of the particle.

Any state of spin is linear superposition of two basic states, and the degree of spin freedom is described in a two-dimensional Hilbert space ϵ_S.

13.1.1   Spin Observable

Let S be the observable vector spin, that is, a set of three observable $S_x, S_{y,z}\, S$. These three observables follow the commuting relations according to,

$$\mathbf{S} \times \mathbf{S} = i\hbar \mathbf{S} \tag{13.1}$$

Each of the observables S_x, S_y and S_z has eigenvalues $h/2$. The observable $S^2 = S_x^2 + S_y^2 + S_z^2$ is proportional to the identity in the spin space ϵ_S with eigenvalue $3\hbar^2/4$.

13.1.2 Representation in a Particular Basis

Let us choose a state basis where S^2 and S_z are diagonal, which we denote $|+\rangle, |-\rangle$:

$$S_z|+\rangle = \frac{\hbar}{2}|+\rangle, \qquad S_z|-\rangle = -\frac{\hbar}{2}|-\rangle, \qquad \mathbf{B}^2|\pm\rangle = \frac{3\hbar^2}{4}|\pm\rangle \quad (13.2)$$

The states $|\pm\rangle$ would be $|j = 1/2, m = \pm 1/2\rangle$.

The action of S_x and S_y on the elements of this basis is written:

$$S_x|+\rangle = \frac{\hbar}{2}|-\rangle, \; S_x|-\rangle = \frac{\hbar}{2}|+\rangle, \qquad\qquad (13.3)$$

$$S_y|+\rangle = \frac{i\hbar}{2}|-\rangle, \; S_x|-\rangle = -\frac{i\hbar}{2}|+\rangle \qquad\qquad (13.4)$$

An arbitrary state of spin $|\Sigma\rangle$ is written:

$$|\Sigma\rangle = \alpha|+\rangle + \beta|-\rangle, \quad \text{with} \quad |\alpha|^2 + |\beta|^2 = 1 \qquad (13.5)$$

The probabilities of finding $+\hbar/2$ and $-\hbar/2$ in a measure of S_z on this state are $P(+\hbar/2) = |\alpha_+|^2$, $P(-\hbar/2) = |\alpha_-|^2$.

13.1.2.1 Matrix Representation

It is convenient to use a matrix representation for the above state vectors and observables:

$$|+\rangle = \begin{pmatrix} 1 \\ 0 \end{pmatrix}, \qquad |-\rangle = \begin{pmatrix} 0 \\ 1 \end{pmatrix}, \qquad |\Sigma\rangle = \begin{pmatrix} \alpha_+ \\ \alpha_- \end{pmatrix} \qquad (13.6)$$

We can use the Pauli matrices $\Sigma \equiv \sigma_x, \sigma_y, \sigma_z$,

$$\sigma_x = \begin{pmatrix} 0 & 1 \\ 1 & 0 \end{pmatrix}, \qquad \sigma_y = \begin{pmatrix} 0 & -i \\ i & 0 \end{pmatrix}, \qquad \sigma_z = \begin{pmatrix} 1 & 0 \\ 0 & -1 \end{pmatrix} \qquad (13.7)$$

Which satisfy the commuting relations:

$$\sigma \times \sigma = 2i\sigma \qquad\qquad (13.8)$$

Spin observables are written in the following way:

$$S = \frac{\hbar}{2}\sigma \qquad (13.9)$$

In this basis, the eigenstates of $|\pm\rangle_x$ of S_x and $|\pm\rangle_y$ of S_y are:

$$|\pm\rangle_x = \frac{1}{\sqrt{2}}\begin{pmatrix} 1 \\ \pm 1 \end{pmatrix}, \qquad |\pm\rangle_y = \frac{1}{\sqrt{2}}\begin{pmatrix} 1 \\ \pm i \end{pmatrix} \qquad (13.10)$$

13.1.2.2 General Spin State

Consider the most general spin state $|\Sigma\rangle$. To a global phase factor that is not of physical importance, this state can always be written:

$$|\Sigma\rangle = e^{-i\varphi/2}\cos(\theta/2)|+\rangle + e^{i\varphi/2}\sin(\theta/2)|-\rangle \qquad (13.11)$$

where $0 \le \theta \le \pi$ and $0 \le \phi < 2\pi$. It is then simple to check that $|\Sigma\rangle$ is eigenstate with the eigenvalue $\hbar/2$ of the operator $S_u = \hat{u}\cdot S$, projection of the spin on the unit vector axis $\hat{u}$, with polar angle θ and azimuth ϕ. We have:

$$u = \sin\theta\cos\varphi u_x + \sin\theta\sin\varphi u_y + \cos\theta u_z \qquad (13.12)$$

Or, in matrix notation:

$$\mathbf{S}_u = \frac{\hbar}{2}\begin{pmatrix} \cos\theta & e^{-i\varphi}\sin\theta \\ e^{i\varphi}\sin\theta & -\cos\theta \end{pmatrix} \qquad (13.13)$$

In other words, for any state $|\Sigma\rangle$ of spin $1/2$, there exists a vector $\hat{u}$ such that $|\Sigma\rangle$ is the state of the operator $\hat{u}\cdot S$, corresponding to the projection of the spin on the direction $\hat{u}$. This remarkable property does not generalize to a higher spin.

13.2 COMPLETE DESCRIPTION OF A $\frac{1}{2}$ SPIN PARTICLE

The spatial state of a spin $1/2$ in three-dimensional space is described on the Hilbert space ϵ_E of a three dimensional square integrable functions $\mathbb{L}^2(\mathbb{R}^3)$, and the spin state in ϵ_S introduced above.

13.2.1 Hilbert Space

The complete Hilbert space is the tensor product of these two spaces:

$$\epsilon_H = \epsilon_{\text{outer}} \otimes \epsilon_{\text{spin}} \tag{13.14}$$

Every element $|\psi\rangle \in \epsilon_H$ is written:

$$|\psi\rangle = |\psi_+\rangle \otimes |+\rangle + |\psi_-\rangle \otimes |-\rangle \tag{13.15}$$

where $|\psi_+\rangle$ and $|\psi_-\rangle$ are elements of ϵ_E.

We note that an observables of space $A_E(x, p, etc.)$ and the observables of spin $B_S(eg S_x, S_y, etc.)$ act in different spaces and commute.

The tensor product of two such observables is defined by:

$$\left(A_{\text{out}} \otimes B_{\text{sp}}\right)\left(|\psi_\sigma\rangle \otimes |\sigma\rangle\right) = \left(A_{\text{out}}|\psi_\sigma\rangle\right) \otimes \left(B_{\text{sp}}|\sigma\rangle\right), \tag{13.16}$$

where $\sigma = \pm$.

13.2.2 Representation of States and Observables

There are several possible representations of the states, accompanied by corresponding representations of the observables, whose use can be more or less convenient according to the problem under consideration.

13.2.2.1 Hybrid Representation

The state is represented by a vector of spin whose components are square integrable functions:

$$\psi_+(\boldsymbol{r}, t)|+\rangle + \psi_-(\boldsymbol{r}, t)|-\rangle \tag{13.17}$$

Let us recall the physical meaning of this representation: $| + (r, t)2d3r(R, t)2D3r)$ is the probability of finding the particle in a neighborhood $D3r$ of the point r with $+h/2$ as projection of its spin along z. An operator in ϵ_E acts on the functions (r, t), a spin operator acts on the following vectors $|+\rangle$ and $|-\rangle$ by (12.2), (12.3) and (12.4) and the products of operators $AB sp$ are comes from (12.13).

13.2.2.2 *Two-Component Wave Function*

The state vector is then represented in the form:

$$\begin{pmatrix} \psi_+(\boldsymbol{r}, t) \\ \psi_-(\boldsymbol{r}, t) \end{pmatrix} \tag{13.18}$$

The physical interpretation of ψ_+ and ψ_- as the probability amplitudes of the pair of variables (r, S_z).

13.2.2.3 *Atomic States*

In many atomic physics problems, it is convenient to use the quantum numbers n, l, m for the classification of the states $|n, l, m\rangle$ which form a basis in ϵ_E. The spin is introduced in the space generated by the family $|n, l, m\rangle \otimes |\sigma\rangle$, where the quantum number of spin takes the two values ± 1. It is convenient to use compact notation:

$$|n, \ell, m, \sigma\rangle \equiv |n, \ell, m\rangle \otimes |\sigma\rangle \tag{13.19}$$

where the states of an electron are described by four quantum numbers. The action of the general operators on the states $|n, l, m, \sigma\rangle$ immediately comes from the considerations developed above.

13.3 MAGNETIC SPIN MOMENT

To the cinematic moment of spin of a particle corresponds a magnetic moment which is proportional to it:

$$\boldsymbol{\mu} = \gamma \, \mathbf{S} = \mu_0 \boldsymbol{\sigma} \tag{13.20}$$

with $\mu_0 = \gamma\hbar/2$. This proportionality is fundamental. It implies the commuting relations between the components of the magnetic moment which are established in the study of the experiment of Stern and Gerlach. Let us recall that this experiment gives direct access to the nature of the kinetic momentum of the atom. In fact, the deflection of the beams is proportional to μ_z, and therefore to J_z. If the magnetic moment of the atom is due to an orbital kinetic momentum, an odd number of spots must be observed. The observation of a split in an even number of spots, two for monovalent atoms such as silver, is a proof of the existence of half-integer momentums.

13.3.1 Abnormal Zeeman Effect

Let us plunge an atom, prepared in a level of energy E and momentum j, in a magnetic field B parallel to the z axis. The potential energy is written:

$$W = -\boldsymbol{\mu} \cdot \mathbf{B} \tag{13.21}$$

The corresponding level is split in $2j + 1$ levels of respective energy:

$$E = \gamma \hbar m B_0 \quad \text{with} \quad m = -j, \dots, j \tag{13.22}$$

There is then a corresponding reduction for each line observed in the spectrum. We know that j must be integral if all the kinetic momentum are orbital or, in other words, classically interpretable momentums. Now, if j is integer, $2j+1$ is odd and we expect a split of each level in an odd number of sub-levels. The study of the split of the atomic lines under the effect and of an external magnetic field was observed for the first time by Zeeman in 1896 [132]. In many cases, and especially for alkaline atoms, the case is that there is a split of the levels into an even number of sub-levels, an abnormal effect.

13.3.2 Magnetic Moment of an Elementary Particle

The electron, the proton and the neutron have a spin $1/2$. The corresponding spin magnetic moment is related to the spin s by the relation $\mu = \gamma s$. The experiments gives the following values of the gyromagnetic ratios γ:

$$\begin{aligned}
\text{electron} \quad & \gamma \approx 2\gamma_0 = -q/m_e \\
\text{proton} \quad & \gamma \approx +2.79\, q/m_p \\
\text{neutron} \quad & \gamma \approx -1.91\, q/m_p
\end{aligned} \tag{13.23}$$

The possible values of the measurement results of a component of these magnetic moments are then:

$$\begin{aligned}
\text{electron} \quad & \mu_z = \pm\mu_B = \mp q\hbar/2m_e \\
\text{proton} \quad & \mu_z = \pm 2.79\, q\hbar/2m_p \\
\text{neutron} \quad & \mu_z = \pm 1.91\, q\hbar/2m_p
\end{aligned} \tag{13.24}$$

The quantity $\mu_B = -9,27410^{-24} JT^{-1}$ is called the Bohr magneton. The quantity $\mu_N = q\hbar/2m_p = 5,05110^{-27} JT^{-1}$ is called a nuclear magnet.

13.4   SPATIAL AND SPIN UNCORRELATED VARIABLES

In most physical situations, the Stern and Gerlach experiment, for example, there are correlations of space and spin variables. Including the variable σ in (12.16) implies that the energy levels of the states $|n, l, m, +\rangle$ and $|n, l, m, -\rangle$ are degenerated. In reality, corrections to this approximation appear, like the fine structure of the hydrogen atom, due to the interaction between the intrinsic magnetic moment of the electron and the electromagnetic field created by the proton. The energy degeneration is then partially lifted and the new eigenstates of the Hamiltonian are linear combinations of the initial states $|n, l, m, \sigma\rangle$. In other words, in a fixed energy level, the spatial wave function of the electron depends on its spin state and the two random variables r and S_z are correlated. However, this correlation is often extremely low. In this case, the two random variables r and S_z can be considered as independent and their probability law is factorized. Such a physical situation will be represented by a factorized state vector:

$$\Phi(r, t) \begin{pmatrix} \alpha_+(t) \\ \alpha_-(t) \end{pmatrix} \qquad (13.25)$$

If spin measurements are made in this case, the results will be independent of the position of the particle. The only observables that are involved are Hermitian matrices 2×2 with numerical coefficients.

13.5   THE MAGNETIC RESONANCE

We have indicated in the preceding paragraph the fundamental relationship between momentum and magnetic moment $\mu = \gamma J$. The determination of the proportionality constant γ is a relevant issue. For fundamental objects like the electron or the proton, its precise measurement constitutes a crucial test of the theories of these particles. For more complex nuclei in molecules, the value of γ provides valuable information about the environment and the chemical bonds involved in these molecules. We will write below how to perform a precise measure of γ. As is often the case in physics, it is a question of taking advantage of a phenomenon of resonance. We shall then return to the applications of this magnetic, electronic or nuclear resonance as the case may be.

13.5.1 Larmor Precession in a Fixed Magnetic Field B_0

Let us choose the z-axis parallel to the field B_0. If we are not concerned with spatial variables, the Hamiltonian writes:

$$H = -\boldsymbol{\mu} \cdot \boldsymbol{B}_0 = -\mu_0 B_0 \sigma_z \tag{13.26}$$

and we can write,

$$-\frac{\mu_0 B_0}{\hbar} = \frac{\omega_0}{2}, \quad \text{let's define} \quad \omega_0 = -\gamma B_0 \tag{13.27}$$

The eigenstates of H are the eigenstates $|+\rangle$ and $|-\rangle$ of σ_z. Consider any state $|\psi(t)\rangle$ such that $|\psi(0)\rangle = \alpha|+\rangle + \beta|-\rangle$ with $|\alpha|^2 + |\beta|^2 = 1$. Its evolution in time is given by:

$$|\psi(t)\rangle = \alpha e^{-i\omega_0 t/2}|+\rangle + \beta e^{i\omega_0 t/2}|-\rangle \tag{13.28}$$

The mean value $\langle \mu \rangle$ is:

$$\langle \mu_x \rangle = 2\mu_0 \operatorname{Re}(\alpha^* \beta e^{i\omega_0 t}) = C \cos(\omega_0 t + \varphi) \tag{13.29}$$

$$\langle \mu_y \rangle = 2\mu_0 \operatorname{Im}(\alpha^* \beta e^{i\omega_0 t}) = C \sin(\omega_0 t + \varphi) \tag{13.30}$$

$$\langle \mu_z \rangle = \mu_0 \left(|\alpha|^2 - |\beta|^2 \right) \tag{13.31}$$

where C and ϕ are the modulus and the phase of the complex number $\alpha * \beta$, respectively. The projection $\langle \mu_z \rangle$ of the magnetic moment on the axis of the field is independent of time and the component of $\langle \mu \rangle$ perpendicular to B rotates at angular velocity ω_0, we found the Larmor precession. The fact that $\langle \mu_z \rangle$ is a constant of motion is a consequence of the commutation relation $[H, \mu_z] = 0$ and the Ehrenfest theorem. A simple method is deduced to measure the angular frequency ω_0. A coil is placed in a plane parallel to B_0 and a macroscopic assembly of spins, all prepared in the same state $|\psi(0)\rangle$. The precession of $\langle \mu \rangle$ at frequency ω_0 causes a periodic variation of the magnetic flux in the coil, hence a measurable induced current of the

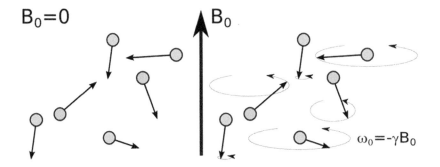

Figure 13.1: Larmor precession [The left shows spins randomly arrange with no magnetic field. Right shows how each spin stars processing with frequency ω_0 around the axis define by the magnetic field, regardless its initial position.].

same frequency is produced. However, this method is less precise than the magnetic resonance technique.

13.5.2 Superposition of a Fixed and a Rotating Field

A technique, developed by Rabi in the 1930's [133], allows a very precise measurement of ω_0 by means of a resonance phenomenon. Let us consider a magnetic moment whose gyromagnetic ratio γ is to be measured in a known field B_0, and superimpose a weak field B_1, rotating in the xy plane at the adjustable angular velocity ω. Such a field can be built of two coils placed along x and y, fed with alternating current of pulsation ω and phase shifts of $\pi/2$. At resonance, that is to say for $\omega = \omega_0$, the spin switches between the two possible states $|\pm\rangle$. The Hamiltonian now has the following form:

$$H = -\boldsymbol{\mu} \cdot \boldsymbol{B} = -\mu_0 B_0 \sigma_z - \mu_0 B_1 \cos \omega t \, \sigma_x - \mu_0 B_1 \sin \omega t \, \sigma_y \quad (13.32)$$

and,

$$|\psi(t)\rangle = a_+(t)|+\rangle + a_-(t)|-\rangle \quad (13.33)$$

The Schrödinger equation leads to the differential system of equations for the coefficients $a_\pm(t)$:

$$i\dot{a}_+ = \frac{\omega_0}{2}a_+ + \frac{\omega_1}{2}e^{-i\omega t}a_- ,$$ (13.34)

$$i\dot{a}_- = \frac{\omega_1}{2}e^{-i\omega t}a_+ - \frac{\omega_0}{2}a_-$$ (13.35)

where we have $\mu_0 B_0/\hbar = -\omega_0/2$, $\mu_0 B_1/\hbar = -\omega_1/2$. The change of function $b_\pm(t) = e^{(\pm i\omega t/2)}a_\pm(t)$ gives:

$$i\dot{b}_+ = -\frac{\omega - \omega_0}{2}b_+ + \frac{\omega_1}{2}b_- ,$$ (13.36)

$$i\dot{b}_- = \frac{\omega_1}{2}b_+ + \frac{\omega - \omega_0}{2}b_-$$ (13.37)

The above transformation is the quantum form of a change of reference that changes from the laboratory frame to the rotating frame at the angular velocity ω about the axis z. We have therefore chosen a base of the Hilbert space which depends on time. In this basis, the Hamiltonian is independent of time:

$$\tilde{H} = \frac{\hbar}{2}\begin{pmatrix} -(\omega - \omega_0) & \omega_1 \\ \omega_1 & \omega - \omega_0 \end{pmatrix} = \frac{\hbar}{2}\omega_1\sigma_x - \frac{\hbar}{2}(\omega - \omega_0)\sigma_z$$ (13.38)

It will be verified that equations (12.32,12.33) result in $\ddot{b}_\pm + (\Omega/2)^2 b_\pm$ with:

$$\Omega^2 = (\omega - \omega_0)^2 + \omega_1^2$$ (13.39)

Suppose that the spin is initially in the state $|+\rangle$, or $b_-(0) = 0$. We then find:

$$b_+(t) = \cos\left(\frac{\Omega t}{2}\right) + i\frac{\omega - \omega_0}{\Omega}\sin\left(\frac{\Omega t}{2}\right)$$ (13.40)

$$b_-(t) = -\frac{i\omega_1}{\Omega}\sin\left(\frac{\Omega t}{2}\right)$$ (13.41)

The probability that a measure of S_z at time t gives the result $-\hbar/2$ is:

$$\mathcal{P}_{(+\to-)}(t) = |\langle -|\psi(t)\rangle|^2 = |a_-(t)|^2 = |b_-(t)|^2$$

$$= \left(\frac{\omega_1}{\Omega}\right)^2 \sin^2\left(\frac{\Omega t}{2}\right)$$ (13.42)

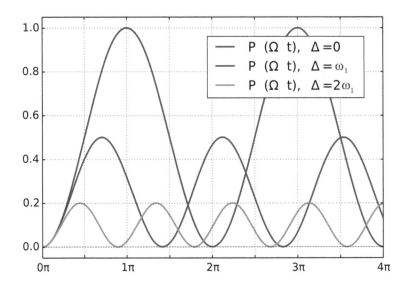

Figure 13.2: Rabi oscillations for several values of $\Delta = \omega - \omega_0$.

This formula, which is due to Rabi [133], clearly demonstrates the resonance phenomenon. There are several cases, if the frequency ω of the rotating field is chosen to be significantly different from the frequency ω_0 then the probability that the spin tilts, that is to say that $S_z = -\hbar/2$ is measured, is very weak for all t. If we choose $\omega = \omega_0$, then $\Omega = \omega_1$ and the spin tilt probability is equal to 1 at times $t_n = (2n + 1)\pi/\omega_1$ with n integer. The amplitude of the rotating field B_1 is very low. Finally for $|\omega - \omega_0| \, \omega_1$, the probability amplitude oscillates with a maximum appreciable amplitude but less than 1.

13.5.3　Rabi's Experiment

The resonance effect understood by Rabi in 1939 provides a very precise method of measuring a magnetic moment. The Rabi apparatus is the combination of two Stern and Gerlach deflectors with their magnetic field in opposite directions (13.3). Between the magnets there is an area where a uniform field B_0 and a rotating field B_1 are superimposed. Consider first the effect and the two Stern and Gerlach magnets in the absence of the fields B_0 and B_1. A particle emitted by the source in the state $|+\rangle$ is deflected upwards, then downwards, and reaches the detector. In the

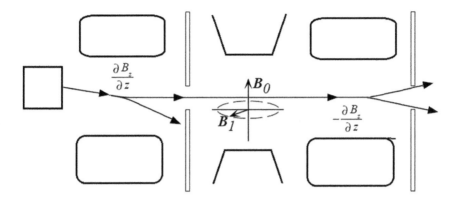

Figure 13.3: Rabi apparatus schematics.

presence of fields B_0 and B_1, this is no longer true. If the frequency ω of the rotating field is close to the Larmor frequency ω_0, the resonance phenomenon changes the μ_z component of the particle. When the spin swings between the two magnets of Stern and Gerlach, the particle is deflected upward by each magnet and is not detected. The signal recorded by the detector as a function of the frequency of the rotating field thus shows a significant decrease when $\omega = \omega_0$ (Figure 13.3). This leads to a measurement of the ratio:

$$\frac{|\mu|}{j} = \frac{\hbar \omega_0}{B_0} \tag{13.43}$$

for a particle of momentum j. This measure is so precise that the determination of B_0 is the main source of error. In practice, as seen in Figure 13.3, the frequency ω is fixed and the field B_0 is varied, or, in a similar manner, the angular frequency ω_0. In 1933, Stern was able to measure the magnetic moment of the proton with a 10% of error [134].

The nuclear magnetic moments are 1000 times lower than magnetic electron moments, we must operate on H_2 or HD molecules, where the effects of the electrons are cancel each other. Thanks to his resonance system, Rabi gained a factor of 1000 in precision: the resonance is selective in frequency, the presence of other magnetic moments is not an issue. Stern observes that he reaches the theoretical limit of precision, fixed by the relations of uncertainty. The great breakthrough of MNR applications came with the work of Felix Bloch at Stanford and Edward Purcell at MIT in 1945 [136, 135]. Bloch and Purcell were able to operate not on molecular beams but directly on condensed matter: We then have a macroscopic number of spins, the signals are much more intense, the experiments much more manageable. The resonance is observed by measuring, for example, the absorption of the wave generating the rotating field B_1. The population difference between the two states $|+\rangle$ and $|-\rangle$, necessary to have a signal, results from the thermodynamic equilibrium. In a field $B_0 = 1T$, the magnetic energy of a proton is $\mu_p B \approx 9 \times 10^{-8}$ eV and the population divergence between the two spin states due to the Boltzmann factor at room temperature is $\pi_+ - {}_- \approx 3 \times 10^{-6}$. This difference is small, but sufficient to observe a reasonable signal because we work with samples containing a macroscopic number of spins, typically 10^{23}.

13.5.4 Applications of Magnetic Resonance

The applications of magnetic resonance are innumerable, in fields as varied as the physics of the solid state and low temperatures, chemistry, biology or medicine. The spin plays the role of local probe within the material. MNR has transformed chemical analysis and the determination of the structure of molecules. It has become a tool of choice in biology and has made considerable progress in the knowledge of macromolecules and, in general, in molecular biology. It has also revolutionized the medical diagnosis and physiology. It allows to measure and visualize in three dimensions and with a spatial precision to the millimeter, the water concentration of the "soft matter," which, unlike the bones, is difficult to observe by X-rays. In this way, the metabolism of living tissues can be detected, internal lesions and tumors can be detected.

13.5.5 Rotation of 2π of a $1/2$ Spin

It seems evident from the common geometric sense that a rotation of 2π of a system around a fixed axis is equivalent to the identity. However, this is not true for a $1/2$ spin. Let us resume the calculation of 5.1 and assume that we are at $t = 0$, where the state of the spin is $|+x\rangle$:

$$|\psi(t = 0)\rangle = \frac{1}{\sqrt{2}}(|+\rangle + |-\rangle) \tag{13.44}$$

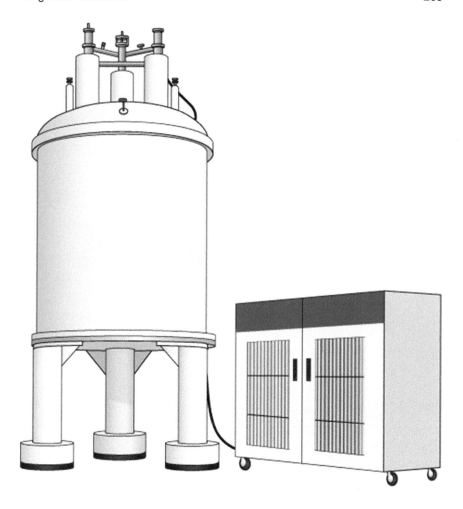

Figure 13.4: Typical synchrotron spectrometer scheme, used as analytical tool in chemistry and biology.

The mean magnetic moment given by (12.25), (12.26) and (12.27) is $\langle \mu \rangle = \mu_0 u_x$. Equation (12.24) gives the evolution of this state. At the time $t = 2\pi/\omega_0$, classically, the system has processed 2π around B. Quantitatively, it is also verified that the magnetic moment has returned to its initial value $\langle \mu \rangle = \mu_0 u_x$. Additionally, we verify that $|\psi(t)\rangle$ is a new eigenstate of S_x with eigenvalue $+\hbar/2$, but, its state vector has changed sign:

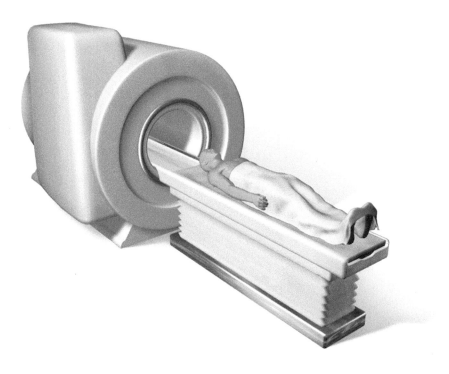

Figure 13.5: Magnetic resonance image device used into medical application for diagnosis.

$$|\psi(t = 2\pi/\omega_0)\rangle = -\frac{1}{\sqrt{2}}(|+\rangle + |-\rangle) = -|\psi(0)\rangle \qquad (13.45)$$

A rotation of 2π is therefore not equivalent to the identity for a spin $1/2$. Only the rotations of multiple integers of 4π return identically to the initial state vector. This property can also be seen as a dependence on $e^{im\phi}$, in the case of orbital kinetic moments: the application of $e^{im\phi}$ for $m = 1/2$ and $\phi = 2$ would give $e^{i\pi} = -1$. This peculiarity constituted a subject of controversy. This phase of the state vector, acquired in a rotation of 2π has a physical meaning. The experimental verification was only given in the 1980's in a series of remarkable experiments. The $1/2$ spins are passed through a two-way interferometer. In one of the channels, a magnetic field effecting a rotation of 2π is arranged. The change of sign of the wave function for the thus modified path results in a displacement of the interference fringes. The fact that a rotation of 2π of a spin $1/2$ is not equivalent to identity, contrary to common sense, shows once again that this magnitude is only quantum.

CHAPTER 14

INTRODUCTORY THEORY OF SCATTERING

The theory of scattering consists in studying the collision also called dispersion, between two or more particles. We will study the simplest case of the collision between two particles without spin. It is assumed that the interaction between the two particles is described by a potential energy which

$$V(\vec{x}_2 - \vec{x}_1) \tag{14.1}$$

where $\vec{x}$ and $\vec{y}$ are the positions of particles 1 and 2. It is assumed that the potential V is short range. If $|\vec{x}_2 - \vec{x}_1|$ is large, we will see that we can use the assumption

$$V(\vec{x}) = o\left(\frac{1}{|\vec{x}|}\right) \tag{14.2}$$

In the study of the problem with two isolated bodies, it is possible to describe the relative motion

$$\vec{x} \equiv \vec{x}_2 - \vec{x}_1 \tag{14.3}$$

in the reference frame of the center of mass and that this relative movement is assumed from the reduce mass $m = \frac{m_1 m_2}{m_1 + m_2}$. The Hamiltonian is then

$$H = \frac{p^2}{2m} + V(\vec{x}) \tag{14.4}$$

For example, for the scattering of an electron on a proton, $m_e \ll m_p$, so the center of mass is placed approximately on the proton $\vec{x}_p$ which is practically immobile at $\vec{x}_p = 0$, and the relative motion $\vec{x}(t) = \vec{x}_e - \vec{x}_p = \vec{x}_e$ describes the motion of the electron. In this case the potential is $V(\vec{x}) = -e^2/4\pi\epsilon_0|\vec{x}|$, which does not scatter fast enough to satisfy the hypothesis (14.2).

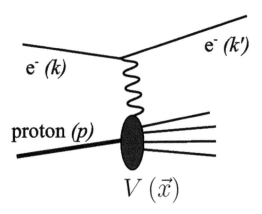

In quantum mechanics, particles are described by waves or wave functions, so the problem consists essentially in describing the scattering of a incident wave by the potential $V(\vec{x})$.

Historically, the discovery of the quantum world and the particles was made by studying collision processes. Many physics experiments are based on collisions, or more precisely of scattering. In 2012 particle physicists announced the long awaited discovery of the Higgs Boson in the collision processes at the Large Hadron Collider (LHC). The following theory holds for the waves in general. It therefore adapts for light, or sound. We will limit ourselves to the study of stationary solutions, called the theory of "stationary scattering." The essential question will be to express the scattered wave component from the incident wave.

14.1 AMPLITUDE OF SCATTERING $F(K, \theta, \phi)$

The following proposition gives a description of the stationary waves with energy E.

If $\psi(\vec{x})$ is a stationary wave with energy $E = \hbar^2 k^2/2m$, that is, verifies $H\psi(\vec{x}) = E\psi(\vec{x})$, then for $r = |\vec{x}| \gg 1$, is written in a unique way:

$$\psi(\vec{x}) = \underbrace{a_-(\theta, \varphi)\frac{e^{-ikr}}{r}}_{\text{incoming wave}} + \underbrace{a_+(\theta, \varphi)\frac{e^{+ikr}}{r}}_{\text{outgoing wave}} + o\left(\frac{1}{r}\right), \qquad \text{where} \quad r = |\vec{x}|$$

(14.5)

where $a_\pm(\theta, \phi)$ is called incoming/outgoing spherical amplitudes, are functions of the angular variables only, and $o(1/r)$ describes a term that decreases faster than $1/r$ for $r \to]\infty$.

We can demonstrate this for $r = |\vec{x}| \gg 1$, away from the collision site, where $V(\vec{x}) \to 0$, so the Hamiltonian approximates to

$$H \simeq H_0 + o\left(\frac{1}{|\vec{x}|}\right) \qquad H_0 = \frac{p^2}{2m} = -\frac{\hbar^2}{2m}\Delta \qquad (14.6)$$

which is that of a free particle. We have

$$H\psi = E\psi \Leftrightarrow -\frac{\hbar^2}{2m}\Delta\psi = \frac{\hbar^2 k^2}{2m}\psi + o\left(\frac{1}{|\vec{x}|}\right)\psi \Leftrightarrow \Delta\psi = -k^2\psi + o\left(\frac{1}{|\vec{x}|}\right)\psi$$

(14.7)

In spherical coordinates, the Laplacian is written

$$\Delta\psi = \frac{1}{r}\frac{\partial^2(r\psi)}{\partial r^2} + \frac{1}{r^2}\left(\frac{1}{\sin\theta}\frac{\partial}{\partial\theta}\left(\sin\theta\frac{\partial\psi}{\partial\theta}\right) + \frac{1}{\sin^2\theta}\frac{\partial^2\psi}{\partial\varphi^2}\right) \qquad (14.8)$$

If we write $\psi(r, \theta, \phi) = R(r)a(\theta, \phi)$, (14.8) is given by

$$\Delta\psi = \frac{a}{r}\frac{d^2(rR)}{dr^2} + o\left(\frac{1}{r}\right) = -k^2 aR \quad \leftrightarrow \quad \frac{d^2(rR)}{dr^2} = -k^2(rR) + o(1)$$

(14.9)

This is the harmonic oscillator equation and has two independent solutions $rR(r) = e^{\pm ikr} + O(1)$, or

$$R(r) = \frac{e^{\pm ikr}}{r} + o\left(\frac{1}{r}\right) \qquad (14.10)$$

In order to illustrate the preceding formula (14.10), consider the particular case of a planar wave of energy E propagating in the direction of the z-axis. A plane wave corresponds to a free particle, i.e., to a situation without potential V. The following proposition gives its decomposition in spherical waves. The result shows that the incoming wave comes only from the direction $\theta = \pi$ and that the outgoing wave, also called transmitted wave, moves in the direction $\theta = 0$.

A plane wave according to z can be written:

$$\psi_{\vec{p}}(\vec{x}) = e^{i\frac{\vec{p}\cdot\vec{x}}{\hbar}} = e^{i\vec{k}\cdot\vec{x}}, \quad \text{with} \quad \vec{p} = \hbar\vec{k} \tag{14.11}$$

$$\hat{H}_0\psi_{\vec{p}} = E\psi_{\vec{p}}, \quad E = \frac{p^2}{2m} = \frac{\hbar^2 k^2}{2m} \tag{14.12}$$

$$e^{ikz} = a_+(\theta)\frac{e^{+ikz}}{r} + a_-(\theta)\frac{e^{-ikz}}{r} + o\left(\frac{1}{r}\right) \tag{14.13}$$

$$\text{with} \quad a_- = \frac{2\pi}{k}\delta(\theta - \pi) \tag{14.14}$$

$$a_+ = -\frac{2\pi}{k}\delta(\theta) \tag{14.15}$$

where δ is the Dirac distribution. More generally, for the free problem, the asymptotic amplitudes a_+ and a_- are connected by the relation:

$$a_+ = -\hat{P}a_- \tag{14.16}$$

where $\mathcal{P} : C(S^2) \to C(S^2)$ is the parity operator defined by $\mathcal{P}a(\theta, \phi) = a(\mathcal{P}(\theta, \phi))$ with the parity transformation into spherical coordinates given by, $\mathcal{P} : (\theta, \phi) \to (\pi - \theta, \phi + \pi)$.

Note that the amplitudes $a_\pm(\theta)$ are independent of ϕ. Let us demonstrate this in spherical coordinates $z = r\cos\theta$ where the plane wave is written:

$$e^{ikz} = e^{ikr\cos\theta} \tag{14.17}$$

Let's simplify this expression in a far field, that is to say, on a sphere of a radius r which is very large $r \gg 1$. This is a function of the form $e^{i\varphi(\theta)}$ with a phase $\varphi(\theta) = kr\cos\theta$. We have also $\varphi'(\theta) = -kr\sin\theta$ which is non-zero unless $\theta = 0, \pi$. This means that this phase function oscillates very quickly for $r \gg 1$. According to the theorem of the non-stationary phase, we deduce that $e^{i\varphi(\theta)}$ is negligible for $\theta \neq 0.\pi$. We still have to calculate the value of e^{ikz} in $\theta = 0, \pi$. For this purpose, the theorem of the stationary phase is used. The small difficulty here is that these two points are badly described by the spherical coordinates. Near these points, we will instead use the coordinates $(\tilde{x}, \tilde{y})$ given by $\tilde{x} = x/r = \cos\phi\sin\theta$ and $\tilde{y} = y/r = \sin\theta\sin\phi$ and we are interested in the neighborhood of the point $(\tilde{x}, \tilde{y}) = (0, 0)$. We then have

$$e^{ikz} \simeq e^{i\varphi} \tag{14.18}$$

with

$$\varphi = kz = k\left(r^2 - x^2 - y^2\right)^{1/2} = kr\left(1 - \tilde{x}^2 - \tilde{y}^2\right)^{1/2}$$

$$\simeq kr\left(1 - \frac{1}{2}\tilde{x}^2 - \frac{1}{2}\tilde{y}^2 + o(\tilde{x}^2, \tilde{y}^2)\right) \tag{14.19}$$

The derivatives are

$$\frac{\partial\varphi}{\partial\tilde{x}} = -kr\tilde{x}, \quad \frac{\partial\varphi}{\partial\tilde{y}} = -kr\tilde{y}, \quad \frac{\partial^2\varphi}{\partial\tilde{x}^2} = \frac{\partial^2\varphi}{\partial\tilde{y}^2} = -kr \tag{14.20}$$

We observe the effect of the stationary phase $\partial\varphi = 0$ on $\tilde{x} = \tilde{y} = 0$. If $f(\tilde{x}, \tilde{y})$ is a nonzero test function close to this point, we have the first order $d\Omega\, d\tilde{x}d\tilde{y}$ and we write the formula of the stationary phase:

$$\int_{S^2} e^{i\varphi} f\, d\Omega = \iint e^{i\varphi(\tilde{x}, \tilde{y})} f\, d\tilde{x}d\tilde{y}$$

$$= e^{i\varphi(0,0)} f(0,0) \left(\frac{-i\sqrt{2\pi}}{\left(\frac{\partial^2\varphi}{\partial\tilde{x}^2}\right)}\right)^{1/2} \left(\frac{-i\sqrt{2\pi}}{\left(\frac{\partial^2\varphi}{\partial\tilde{y}^2}\right)}\right)^{1/2}$$

$$= e^{ikr} f(0,0) \frac{(-2\pi)}{kr} \tag{14.21}$$

So, close to the point $\theta = 0$,

$$e^{ikz} \equiv e^{ikr}\delta(\theta)\frac{(-2\pi)}{kr} = \frac{e^{ikr}}{r}a_+(\theta) \tag{14.22}$$

With $a_+(\theta) = (-2\pi)/k\delta(\theta)$. The same calculation is made in the vicinity of the point $\theta = \pi$ to find $a_-(\theta)$.

An incoming spherical wave does not correspond to an usual experimental situation. In an experimental situation, the incoming wave is more often modeled by an incident plane wave. This leads to the following definition:

In the theory of steady-state scattering, a solution of the equation $H\psi = E\psi$, with energy $E = \hbar^2 k^2/2m$, such as in far field $r \gg 1$ is

$$\psi(\vec{x}) = \underbrace{e^{ikz}}_{\psi_i} + \underbrace{(k, \theta, \varphi)\left(\frac{e^{ikr}}{r}\right)}_{\psi_s} + o\left(\frac{1}{r}\right) \tag{14.23}$$

Formed of an incident wave and transmitted one along the z axis, denoted ψ_i and a spherical wave scattered, denoted ψ_s. The function $f(k, \theta, \phi)$ is called amplitude of scattering and is unique.

We also denote $\vec{k}' = k\vec{x}/r$ the wave vector scattered in the direction (θ, ϕ). Then f is a function of $\vec{k}'$:

$$f(\vec{k}') \equiv f(k, \theta, \varphi) \tag{14.24}$$

According to (14.13), the spherical wave amplitudes (14.23) are

$$a_-(\theta) = -\frac{2\pi}{k}\delta(\theta - \pi) \quad a_+(\theta, \varphi) = \frac{2\pi}{k}\delta(\theta) + f(k, \theta, \varphi) \tag{14.25}$$

The term $\delta(\theta)$ corresponds to the forward wave transmitted.

Let us recall some aspects of the probability current.

For any quantum wave $\psi(\vec{x}, t)$, the probability density is given by

$$P(\vec{x}, t) = |\psi(\vec{x}, t)|^2 \tag{14.26}$$

We have the probability conservation equation

$$\frac{\partial P}{\partial t} + \nabla \cdot \vec{j} = 0 \tag{14.27}$$

where the probability current density is determined by

$$\vec{j}(\vec{x}, t) = \Re\left(\bar{\psi}(\vec{x}, t)\left(\hat{\vec{v}}\psi(\vec{x}, t)\right)\right) \tag{14.28}$$

and $\vec{v} = \vec{p}/m = -i\hbar/m\vec{\nabla}$ is the velocity operator.

A demonstration is given by assuming $V(\vec{x})$ real,

$$\frac{\partial P}{\partial t} = 2\,\Re(\bar{\psi}\partial_t\psi)$$

$$= 2\,\Re\left(\bar{\psi}(\vec{x})\left(-\frac{i}{\hbar}\hat{H}\psi\right)(\vec{x})\right)$$

$$= 2\,\Re\left(\bar{\psi}(\vec{x})\left(-\frac{i\hbar}{2m}(-\nabla^2)\psi\right)(\vec{x})\right)$$

$$= -\nabla\Re\left(\bar{\psi}(\vec{x})\left(-\frac{i\hbar}{m}(\nabla)\psi\right)(\vec{x})\right)$$

$$= -\nabla \cdot \vec{j} \tag{14.29}$$

According to (14.28) the vector field $\vec{j}(\vec{x}, t)$ characterizes the displacement of the probability density of the wave function ψ. If $d^2\vec{s}$ is a surface element then $d^2\vec{s}\cdot\vec{j}$? is the probability of crossing this area per unit time. To comment (14.28), one also recalls the transport equation. If $\phi_t : \mathbb{R}^3 \to \mathbb{R}^3$ is the flux at time t generated by the vector field $\vec{j}$, if V_0 is a volume that evolves with time t in $V_t = \phi_t(V_0)$ and if $Q_t = \int_{V_t} P(\vec{x}, t)d^3\vec{x}$ is the probability of finding the particle in the volume V_t in the instant t then,

$$\frac{dQ_t}{dt} = \int_{V_t} \left(\frac{\partial P}{\partial t} + \nabla \cdot \vec{j} \right) d^3\vec{x} \tag{14.30}$$

Thus $\frac{dQ_t}{dt} = 0$ for any volume V_0 is equivalent to (14.28).

For incident waves $\psi_i = e^{ikz}$ and scattered $\psi_s = f(k, \theta, \phi))$ intervening in (14.23), the probability currents are respectively:

$$\vec{j}_i(\vec{x}) = \frac{\hbar}{m}\vec{k} \qquad \vec{j}_s(\vec{x}) = \frac{\hbar}{m}\frac{|f(\vec{k}')|^2}{r^2}k\vec{u}_r \tag{14.31}$$

The total current scattered on the sphere of radius r is

$$\underbrace{(4\pi r^2)}_{\text{surface}} j_s = 4\pi \frac{\hbar}{m}|f(\vec{k}')|^2 k \tag{14.32}$$

is independent of r.

This can be seen in Cartesian coordinates, writing $\vec{\nabla}\psi_s = ike^{ikz}\hat{z}$, where

$$\begin{aligned}
\vec{j}_i(\vec{x}) &= \mathfrak{R}\left(\bar{\psi}_i\left(-\frac{i\hbar}{m}(\nabla\psi_i) \right) \right) \\
&= \frac{\hbar}{m}(0, 0, k) \\
&= \frac{\hbar}{m}\vec{k}
\end{aligned} \tag{14.33}$$

In spherical coordinates the gradient is

$$\nabla\psi_{\text{scatt}} = \left(\frac{\partial\psi}{\partial r}, \frac{1}{r}\frac{\partial\psi}{\partial\theta}, \frac{1}{r\sin\theta}\frac{\partial\psi}{\partial\varphi} \right) \tag{14.34}$$

Therefore, and keeping only the real terms

$$\vec{j}_{\text{scatt}}(\vec{x}) = \Re \left(\bar{\psi}_{\text{scatt}} \left(-\frac{i\hbar}{m} \left(\nabla \psi_{\text{scatt}} \right) \right) \right)$$

$$= \frac{\hbar}{m} (k \, |\psi_{\text{scatt}}|^2 \, , \, 0 \, , \, 0)$$

$$= \frac{\hbar}{m} \frac{|f(\vec{k}')|^2}{r^2} k \vec{u}_r \tag{14.35}$$

The differential cross section is

$$\frac{d\sigma}{d\Omega}(\vec{k}') = \frac{\left(r^2 |\vec{j}_s| \right)}{|\vec{j}_i|} = |f(\vec{k}')|^2 \tag{14.36}$$

The total cross section is

$$\sigma_{\text{T}} = \int_{S^2} \left(\frac{d\sigma}{d\Omega} \right) d\Omega \tag{14.37}$$

According to this statement, this is the probability to scatter in the surface element $r^2 d\Omega$, normalized by the incident current probability $|\vec{j}_i|$. According to (14.36), $\frac{d\sigma}{d\Omega}$ and σ_T have units of a surface. The interpretation is that the potential V has a scatter effect as an effectively equivalent surface σ_T and similarly $d\sigma$ is the equivalent surface which would scatter in the solid angle $d\Omega$.

14.2   BORN APPROXIMATION

The above description is very general and applies to any potential V, which decreases for $r \to \infty$ but can be arbitrary near the origin. In general, it is very difficult to calculate the solutions of the stationary Schrödinger equation, and in particular the amplitude of scattering f of Eq. (7.2.7). However, in the absence of potential, $V = 0$, there is no scattering, $f = 0$, so for a very low potential, one can calculate the scattering amplitude, using stationary perturbation theory. The result is called the Born approximation.

For a low potential V, at first order in the perturbation theory, the amplitude is

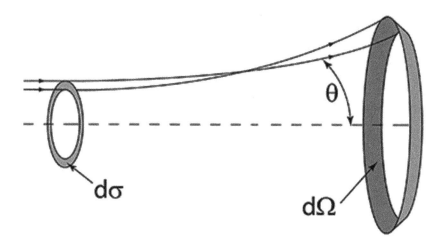

Figure 14.1: Differential cross section.

$$f(\vec{k}') \simeq -\frac{1}{4\pi} \int_{\mathbb{R}^3} e^{i(\vec{k}-\vec{k}')\cdot\vec{x}'} U(\vec{x}') \, d^3\vec{x}' \qquad (14.38)$$

with $U(\vec{x}) = 2m/\hbar^2 V(\vec{x})$ with $\vec{k}' = (k, \theta, \phi)$ the dispersed wave vector.
In fact $f(\vec{k}')$ takes an even simpler form for potentials with spherical symmetry $U(r)$, called central potentials. We want to solve

$$\hat{H}\psi = E\psi \qquad (14.39)$$

With $\hat{H} = \hat{H}_0 + V$, $\hat{H}_0 = \hat{p}^2/2m$. We know that the plane wave $\psi_i = e^{ikz}$ is a solution of $\hat{H}_0\psi_i = E\psi_i$. So

$$\left(\hat{H} - E\right)\psi + \left(\hat{H}_0 - E\right)\psi_{\text{in}} = 0$$
$$\Leftrightarrow V\psi + \left(\hat{H}_0 - E\right)(\psi - \psi_{\text{in}}) = 0$$
$$\Leftrightarrow \psi = \psi_{\text{in}} - \left(\hat{H}_0 - E\right)^{-1} V\psi \qquad (14.40)$$

The latter is called the Lipmann-Schwinger equation, we used the inverse of the operator $\left(\hat{H}_0 - E\right)$ which is not invertible. In fact it is invertible For $z \in (\mathbb{C}/\mathbb{R}^+)$, and here we consider the limit $z \to E \in \mathbb{R}^+$. In a first approximation we have $\psi^{(}0) = \psi_i$. Then to first order we have the expression of ψ in the Born approximation:

$$\psi^{(1)} = \psi_{\text{in}} - \left(\hat{H}_0 - E\right)^{-1} V\psi^{(0)}$$

$$= \psi_{\text{in}} - \left(\hat{H}_0 - E\right)^{-1} V\psi_{\text{in}} \tag{14.41}$$

This operation can be continued to obtain the approximation of order $n : \psi^{(}n) = \psi_i - \left(\hat{H}_0 - E\right)^{-1} V\psi^{(n-1)}$, but there is no guarantee that this sequence converges. The Schwartz kernel of the resolvent, called the Green function of $\hat{H}_0$, is

$$\lim_{z \to E} \langle \vec{x} | \left(\hat{H}_0 - z\right)^{-1} | \vec{x}' \rangle = \left(\frac{\hbar^2}{2m}\right) \frac{e^{\pm ik|\vec{x}-\vec{x}'|}}{4\pi|\vec{x} - \vec{x}'|} \tag{14.42}$$

The signs $\pm$ depends on the limit used to approach $E \in \mathbb{R}$. So with the Born approximation (14.41):

$$\psi^{(1)}(\vec{x}) = \psi_{\text{in}}(\vec{x}) - \int \langle \vec{x} | \left(\hat{H}_0 - z\right)^{-1} | \vec{x}' \rangle V(\vec{x}')\langle \vec{x}' | \psi_{\text{in}} \rangle d^3\vec{x}'$$

$$= e^{ikz} - \int \frac{e^{\pm ik|\vec{x}-\vec{x}'|}}{4\pi|\vec{x} - \vec{x}'|} U(\vec{x}')e^{ikz'} d^3\vec{x}' \tag{14.43}$$

In the far field, $r = |\vec{x}| \gg |\vec{x}'|$ then

$$|\vec{x} - \vec{x}'|^2 = \vec{x}^2 - 2\vec{x} \cdot \vec{x}' + \vec{x}'^2 = r^2\left(1 - \frac{2\vec{x} \cdot \vec{x}'}{r^2} + \frac{\vec{x}'^2}{r^2}\right) \tag{14.44}$$

therefore, as $(1+x)^{1/2} = 1 + 1/2x + O(x)$

$$|\vec{x} - \vec{x}'| = r\left(1 - \frac{\vec{x} \cdot \vec{x}'}{r^2}\right) + O\left(\frac{\vec{x}'^2}{r}\right) = r + O(1) \tag{14.45}$$

So

$$e^{\pm ik|\vec{x}-\vec{x}'|} \simeq e^{\pm ikr}e^{\mp ik\frac{\vec{x}\cdot\vec{x}'}{r}} = e^{\pm ikr}e^{\mp i\vec{k}'\cdot\vec{x}'}, \qquad \vec{k}' = k\frac{\vec{x}}{r} \tag{14.46}$$

Thus, since $e^{ikz'} = e^{i\vec{k}\cdot\vec{x}'}$,

$$\psi^{(1)}(\vec{x}) \simeq e^{ikz} - \frac{e^{\pm ikr}}{4\pi r} \int e^{\mp i\vec{k}'\cdot\vec{x}'} U(\vec{x}') e^{i\vec{k}\cdot\vec{x}'} d^3\vec{x}'$$

$$= e^{ikz} + f(k,\theta,\varphi) \left(\frac{e^{ikr}}{r} \right) \qquad (14.47)$$

where the outgoing solution e^{+ikr} was kept and

$$f(k,\theta,\varphi) = -\frac{1}{4\pi} \int e^{i\vec{\Delta}\vec{x}'} U(\vec{x}') d^3\vec{x}', \qquad \vec{\Delta} = \vec{k} - \vec{k}' \qquad (14.48)$$

The result of Born (14.38) shows that at the first order, the amplitude of scattering $f(\vec{k}')$ is directly connected to the Fourier transform of the potential $U(\vec{x})$. According to the theory of Fourier transforms, it is therefore possible to reconstruct the potential by an inverse Fourier transform. This is called the reverse problem. For example, if the potential is periodic, as in the case in crystallography, where there is scattering of a neutron quantum wave or an electromagnetic wave on a crystal, then the amplitude of the dispersion is the Fourier transforms of a periodic function, and according to the Fourier series theory, it is a periodic lattice of Dirac distributions $\delta(\vec{k}_n)$ called the reciprocal lattice. We thus find the divergence theory of Bragg.

14.3 THE SCATTERING OPERATOR, THE S MATRIX

Let us consider the general problem of the dispersion without making the Born approximation.

We recall a simple but fundamental result that we obtained which is the decomposition (14.5) of a stationary wave in incoming and outgoing amplitudes $a_\pm(\theta, \phi)$, in $r \gg 1$. The following result shows that these two amplitudes are connected to each another.

At $k > 0$, the amplitude of the outgoing wave $a_+(\theta, \phi)$ is expressed from the amplitude $a_-(\theta, \phi)$ by a unit operator called scattering operator or S matrix.

$$\hat{S}(k) : \begin{cases} L^2(S^2) & \longrightarrow & L^2(S^2) \\ a_-(\theta,\varphi) & \longrightarrow & a_+(\theta,\varphi) = \left(\hat{S}(k)\, a_- \right)(\theta,\varphi) \end{cases} \qquad (14.49)$$

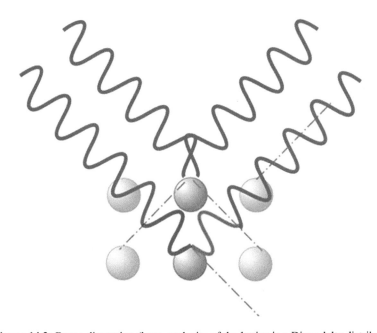

Figure 14.2: Bragg dispersion (here, each site of the lattice is a Dirac delta distribution).

The scalar product on $L^2(S^2)$ is $\int\int_{S^2} \widehat{a}(\theta,\phi)b(\theta,\phi)d\Omega$, with $d\Omega = \sin\theta d\theta d\phi$, the unitarity of the operator is thus written:

$$\|a_+\| = \left\|\hat{S}(k)a_-\right\| = \|a_-\|$$

$$\Longleftrightarrow \iint_{S^2} |a_+(\theta,\varphi)|^2 d\Omega = \iint_{S^2} |a_-(\theta,\varphi)|^2 d\Omega \qquad (14.50)$$

The unitarity property thus represents the conservation of the probability. In the case of a free particle, that is to say with a zero potential $V = 0$, we have seen from (7.2.6) that the scattering operator is expressed from the parity operator:

$$\hat{S}(k) = -\hat{P} \qquad (14.51)$$

In general for a non-zero V, according to the expression (14.25), the amplitude of scattering $f(k,\theta,\phi)$ can be expressed in terms of the matrix S by:

$$f(k,\theta,\varphi) = -\frac{2\pi}{k}\left(\hat{S}(k)\delta(\theta-\pi)\right) - \frac{2\pi}{k}\delta(\theta) \qquad (14.52)$$

Knowing the potential $V(\vec{x})$ which determines the collision process, the problem of general theory of scattering consists in finding the expression of this operator $\hat{S}$. Experimentally, we send an incident wave a_- and observe the outgoing wave a_+. More precisely, the probability distribution $|a_+(\theta, \phi)|^2$ is detected. Hence, we have access to the operator $\hat{S}(k)$. An important problem is to find the forces modeled by the potential $V(\vec{x})$, responsible for the collision; in other words, an important problem is the so-called "inverse problem." Determine the potential $V(\vec{x})$, from the knowledge of the operator $\hat{S}(k)$.

By definition of a unit operator, we can show the conservation of probability, Eq. (14.50). We use the relation (14.27). Consider the sphere S_r of radius $r \gg 1$. Let B_r be the ball of radius r. The probability conservation Eq. (14.27) integrated on the ball B_r, and taking into account that $\frac{\partial P}{\partial t} = 0$ gives

$$0 = \int_{B_r} \nabla \cdot \vec{j} = \int_{S_r} \vec{j} \cdot d^2 \vec{S} \qquad (14.53)$$

Now, according to (14.23) and (14.31) the current has two components (incoming and outgoing)

$$\vec{j} = \vec{j}_+ + \vec{j}_- , \qquad \text{with} \quad \vec{j}_\pm = \pm \frac{\hbar}{m} \frac{|a_\pm(\vec{k}')|^2}{r^2} k \vec{u}_r \qquad (14.54)$$

We have $\int_{S_r} \vec{j}_\pm \cdot d^2\vec{s} = \pm \frac{\hbar}{m} k \int_{S_r} \left| a_\pm(\vec{k}') \right|^2 d\Omega$, therefore the conservation of the probability (14.53) gives

$$\iint_{S^2} |a_+(\theta, \varphi)|^2 d\Omega = \iint_{S^2} |a_-(\theta, \varphi)|^2 d\Omega \qquad (14.55)$$

showing that $\hat{S}(k)$ is unitary.

14.4 PARTIAL WAVES THEORY FOR CENTRAL POTENTIALS

We now assume a simplistic hypothesis, that the potential $V(\vec{x})$ is invariant under rotations, hence is of the form $V(r)$, $r = |\vec{x}|$. This is a central or potential or a spherically symmetric potential.

According to the theory of spherical harmonics, under the action of the group of rotations, the space $L^2(S^2)$ is decomposed into spaces of irreducible representations:

$$L^2(S^2) = \bigoplus_{l=0}^{\infty} \mathcal{D}_l \qquad (14.56)$$

and $Y_{l,m}(\theta, \phi)$, $m = -l ... + l$ and l fixed, form a basis of the space $\mathbb{D}_l$. Hence the functions $Y_{l,m}(\theta, \phi)$ with $l = 0, 1, 2, ... \infty$, $m = -l, -l + 1, ..., +l$ form a basis in the space of functions $a(\theta, \phi) \in L^2(S^2 2)$. Consequently, in this basis, the scattering operator $\hat{S}(k)$ can be expressed by a matrix whose elements are:

$$(S(k))_{(l',m'),(l,m)} = \langle Y_{l',m'} | \hat{S}(k) Y_{l,m} \rangle \qquad (14.57)$$

We will see that for the potentials V with spherical symmetry, this matrix is very Simple: the scattering of a central potential is done independently for each angular moment l, i.e., in each space $\mathcal{D}_l$. This is called a "dispersion channel." In addition, in each channel l, the scattering matrix is characterized by a single phase. More precisely, if $V(r)$ is a spherically symmetric potential then the operator $\hat{S}(k)$ is of the very simple form:

$$\hat{S}(k) = \bigoplus_{l=0}^{\infty} e^{is_l(k)} \hat{Id}_{\mathcal{D}_l} \qquad (14.58)$$

that is to say that it is diagonal in the base $Y_{l,m}$,

$$\langle Y_{l',m'} | \hat{S}(k) Y_{l,m} \rangle = e^{is_l(k)} \delta_{l',l} \delta_{m',m} \qquad (14.59)$$

with amplitude $e^{is_l(k)}$ which depends on l only. In the case $V = 0$, one has $e^{is_l(k)} = (-1)^{l+1}$. For this reason, it is written in general:

$$e^{is_l(k)} = (-1)^{l+1} e^{2i\delta_l} \qquad (14.60)$$

where δ_l is called the phase shift.   .

14.4.1 Heuristic Argument Using the Uncertainty Principle

We have $\delta_l \simeq 0$ if $l \gg k$. In particular at low energy, $k \to 0$, only the mode $l = 0$ is relevant. This is an isotropic dispersion.

If $V(r)$ is a spherically symmetric potential, the effective cross section (14.37) can be written as:

$$\sigma_T = \sum_{l=0}^{\infty} \sigma_l \tag{14.61}$$

With the cross section of the waves given by:

$$\sigma_l = \frac{4\pi(2l+1)}{k^2} \sin^2 \delta_l \tag{14.62}$$

This can be seen if as in (14.52), that the amplitude of scattering is given by

$$f(k,\theta) = -\frac{2\pi}{k}\left(\hat{S}(k)\delta(\theta-\pi)\right) - \frac{2\pi}{k}\delta(\theta) \tag{14.63}$$

independent of ϕ, spherical symmetric. Let us denote $\hat{P}_l : L^2(S^2) \to \mathcal{D}_l$ the orthogonal projector on the space $\mathcal{D}_l$. The closure relation is written $\hat{I}_{L^2(S^2)} = \oplus_l \hat{P}_l$. In the case of spherical symmetry we have $\hat{P}_l\hat{S}(k) = e^{is_l(k)}\hat{P}_l$ then

$$f(k,\theta) = -\frac{2\pi}{k}\sum_l \hat{P}_l\left(\left(\hat{S}(k)\delta_{\theta=\pi}\right) + \delta_{\theta=0}\right)$$

$$= -\frac{2\pi}{k}\sum_l \left(e^{is_l}\hat{P}_l\delta_{\theta=\pi} + \hat{P}_l\delta_{\theta=0}\right) \tag{14.64}$$

By using parity, $\left(\hat{P}_l\delta_{\theta=\pi}\right)(\theta) = (-1)^l\left(\hat{P}_l\delta_{\theta=0}\right)(\theta)$. Moreover $\delta_{\theta=0}$ is independent of ϕ therefore is projected only on $m = 0$:

$$\left(\hat{P}_l\delta_{\theta=0}\right)(\theta) = \langle\theta,\varphi|Y_{l0}\rangle\langle Y_{l0}|0,0\rangle = Y_{l0}(\theta,\varphi)\overline{Y_{l0}(0,0)} \tag{14.65}$$

Therefore,

$$f(k,\theta) = -\frac{2\pi}{k}\sum_l \left(e^{is_l}(-1)^l + 1\right)Y_{l0}(\theta,\varphi)\overline{Y_{l0}(0,0)} \tag{14.66}$$

Finally, according to (14.37) the total effective cross section is

$$\sigma_T = \int_{S^2}|f(k,\theta)|^2 d\Omega = \frac{4\pi^2}{k^2}\sum_l \left|e^{is_l}(-1)^l + 1\right|^2 |Y_{l0}(0,0)|^2 \tag{14.67}$$

To eliminate non-diagonal terms $\sum_{l' \neq l}$, we use the orthogonality relation of the spherical harmonics $\langle Y_{l',m'} | Y_{l,m} \rangle = \delta_{l',l} \delta_{m',m}$. We use $|Y_{l0}(0)| = ((2l+1)/4\pi)^{1/2}$ and

$$\left| e^{i\delta_l}(-1)^l + 1 \right|^2 = \left| -e^{2i\delta_l} + 1 \right|^2 = \left| e^{-i\delta_l} - e^{i\delta_l} \right|^2 = 4\sin^2 \delta_l \quad (14.68)$$

Therefore,

$$\sigma_T = \frac{4\pi^2}{k^2} \sum_l (4\sin^2 \delta_l) \left(\frac{2l+1}{4\pi} \right) = \frac{4\pi}{k^2} \sum_l (2l+1)\sin^2 \delta_l \quad (14.69)$$

CHAPTER 15

INTRODUCTION TO QUANTUM HALL EFFECT

In 1879, Edwin Hall [137], observed for the first time an effect on some materials that today is called Hall effect. This essentially consists in establishing a potential difference, called Hall V_H potential, on the opposite faces of a metal sample and obtain a separation between holes and electrons.

15.1 CLASSICAL HALL EFFECT

The Hall effect is observed in experimentally in thin bands of conductive or semiconductor metal material, driven by an intensity current I subjected to the action of a magnetic field B. To describe the phenomenon we assume the current to be on the x axis and suppose a uniform magnetic field along the z axis: under these conditions there is a potential difference transverse, or Hall voltage V_H, between the lateral surfaces of the conductor.

Starting from this tension, it is possible to determine the cross resistance, which is the Hall resistance, that is equal to:

$$R_H = \frac{V_H}{I} = \frac{B}{n_e e} \tag{15.1}$$

where e is the charge of the electron and n_e the density of the charge carriers. What makes this effect important is that from the applied V_H

voltage it is possible to determine the charge carrier sign. Another important aspect is that the cross-resistance R_H, although having resistance dimensions, does not correspond to the strength resistance of the material. It is also independent of the geometry of the system and is proportional to the intensity of the applied magnetic field.

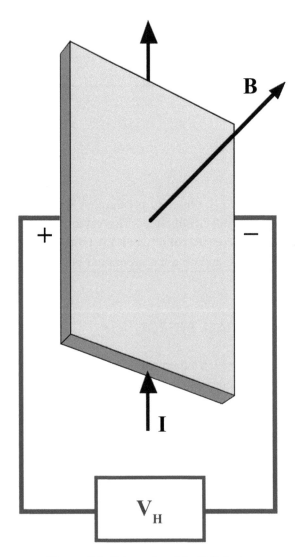

Figure 15.1: Schematics of the Hall device.

15.1.1 The Drude Model

To describe the quantum case it is important first to define some fundamental quantities such as longitudinal resistivity and Hall conductivity.

Let us consider a two-dimensional electronic system: it is observed that in the absence of a magnetic field, the current induced by the electric field E is parallel to the same field and the current density is equal to $j = \sigma E$ where σ is the conductivity of the material. In the presence of a magnetic field B, a transverse current density is created which tends to accumulate charges on the edges of the sample until the balance is reached between the electric force generated by this charge distribution and the Lorentz force generated by the magnetic field . In the ideal case only the B perpendicular component of the 2-d system influences the electronic motion, while only the electric field component lying on the plane of the system has effects on the current. In this case we can then write the relation between the current and the field E by introducing the two-dimensional current density j and the conductivity tensor σ:

$$J = \sigma\, E \tag{15.2}$$

Considering a 2-d system lying on xy plane we can write more explicitly from (15.2):

$$J_x = \sigma_{xx} E_x + \sigma_{xy} E_y , \qquad J_y = \sigma_{yx} E_x + \sigma_{yy} E_y \tag{15.3}$$

Due to the isotropy of the system $\sigma_{xx} = \sigma_{yy}$, called longitudinal conductivity and $\sigma_{xy} = -\sigma_{yx}$, called Hall conductivity. The conductivity is, however, inverse in the matrix sense of the resistivity: $\sigma_{ij} = (\rho^{-1})_{ij}$. So we can rewrite the (15.2) as:

$$E = \rho J \tag{15.4}$$

and the relations (15.3) depending on the resistivity tensor as:

$$E_x = \rho_{xx} J_x + \rho_{xy} J_y , \qquad E_y = \rho_{yx} J_x + \rho_{yy} J_y \tag{15.5}$$

where

$$\rho_L = \rho_{xx} = \rho_{yy} = \frac{\sigma_{xx}}{\sigma_{xx}^2 + \sigma_{xy}^2} \tag{15.6}$$

is called longitudinal resistivity, and

$$\rho_H = \rho_{xy} = -\rho_{yx} = -\frac{\sigma_{xy}}{\sigma_{xx}^2 + \sigma_{xy}^2} \tag{15.7}$$

is called Hall's resistivity.

Once we define these dimensions, we see how they are related to the material's characteristics. Let us consider a Hall bar subjected to the action of an electric field E and a magnetic field B along the z axis.

Under these conditions, the equation of the motion of electrons moving in the plane with a moment $p = mev = (p_x, p_y)$ subjected to the action of the fields is described by the Model of Drude and is given by:

$$m_e \frac{dv}{dt} = -e\left(E + \frac{v \times B}{c}\right) - m_e \frac{v}{\tau} \tag{15.8}$$

where with τ is the relaxation time defined as the average time between

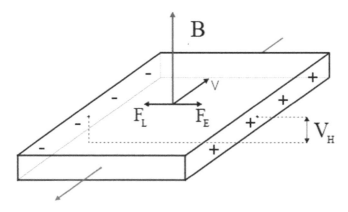

Figure 15.2: The forces acting in the Hall device—a separation of the positive and negative carriers.

successive collisions. The last term indicates a viscous force proportional to the velocity that takes into account the impurities of the material and the interactions between the electrons. Under stationary conditions where $\frac{dv}{dt} = 0$ we can express (1.8) with the following relation:

$$v + \frac{e\tau}{m_e}(v \times B) = -\frac{e\tau}{m_e}E \tag{15.9}$$

The current density is related to the speed from the relation:

$$J = -n_e e v \tag{15.10}$$

where $n_e = N/L_y L_x$ is the two-dimensional electronic density. Using matrix notation, we can rewrite the previous relationships as follows:

$$J \begin{pmatrix} 1 & \omega_c \\ -\omega_c & 1 \end{pmatrix} = \frac{n_e e^2 \tau}{m_e} E \qquad (15.11)$$

where with ω_c is the cyclotron frequency, equal to eB/m_e. In the absence of a magnetic field, the relation $E = \rho j$ holds, where the resistivity is:

$$\rho = \frac{m_e}{n_e e^2 \tau} \qquad (15.12)$$

By comparing (15.11) with the relation (15.5), we find:

$$\rho_L \equiv \rho_{xx} = \frac{m_e}{n_e e^2 \tau} \qquad (15.13)$$

$$\rho_H = \rho_{xy} = \frac{B}{n_e ec} \qquad (15.14)$$

Hall's resistivity is therefore directly proportional to the intensity of the magnetic field in which the system is immersed and is independent of the time of relaxation. This means that it is not affected by the impurities of the material responsible for the scattering between electrons and the material lattice.

15.1.2   Hall Resistance and Resistance

A very important and interesting property concerns the relationship between resistivity and transverse resistance or Hall resistance. Resistivity and resistance are related by the second Ohm's law $R = \rho L/S$ where L is the length of the conductor and S is the transverse section, so they differ by a geometric factor. However, in the case of Hall resistance R_H it is shown that the two coincide.

To prove this we consider a sample of material with length L_y along the y direction, where the V_H has been established so that the current is in the x direction. The Hall resistance measured is, remembering the relation (1.4), equal to

$$R_H = \frac{V_H}{I_x} = \frac{L_y E_y}{L_y J_x} = \frac{E_y}{J_x} = \rho_H \qquad (15.15)$$

So from (15.14) we can write:

$$R_H \equiv \rho_H = \frac{B}{n_e ec} \tag{15.16}$$

It is important to note that Hall's resistance is independent of system geometry but is tied to the type of material given the two-dimensional electronic density. Longitudinal resistance and resistivity are instead related by a dimensionless factor, that takes into account the geometry of the system for which:

$$R_L \equiv \rho_{xx} \frac{L_x}{L_y} = \frac{L_x m_e}{L_y n_e e^2 \tau} \tag{15.17}$$

15.1.3   Finding the Quantum Hall Effect – QHE

The classical Hall effect is observed in the presence of weak magnetic fields. In the case of intense magnetic fields and at low temperatures it is necessary to use quantum mechanics.

The QHE was first discovered experimentally and only later was explained the theory behind the phenomenon.

The first measurements of conductivity and resistivity in specimens subjected to intense magnetic fields were performed in 1978 by Wakabayashi and Kawaji on a Si MOS system [138]. They found that in some regions the longitudinal conductivity σ_{xx} became very small ($\sim 10 - 4$ mho) while Hall's conductivity σ_{xy} behaves typically as $n_e e / B$ corresponding to the inverse of (15.16).

However, the QHE was discovered a few years later in 1980 by physicist Klaus von Klitzing [139]. Making more accurate measurements of R_H resistance, varying the electronic density in a field magnetically fixed at low temperature, he found that in some regions called plateau the resistance of Hall exhibited a constant behavior while the longitudinal resistance was canceled. More precisely he found that Hall's resistance in these regions was exactly the same as:

$$R_H = \frac{1}{\sigma_H} = \frac{h}{ne^2} \tag{15.18}$$

where $\hbar$ is the Plank constant and n an integer $\in \mathbb{N}$ that is determined in experiments with a high precision: about one part per billion.

In other words, what was observed is that resistance and conductivity only depend on fundamental constants and not of the geometry of the system, and that for certain ranges of the magnetic field applied these are quantized.

Quantization has the effect of resetting the probability of dissipative bumps between the charge carriers and the crystalline lattice, hence the longitudinal resistance becomes zero.

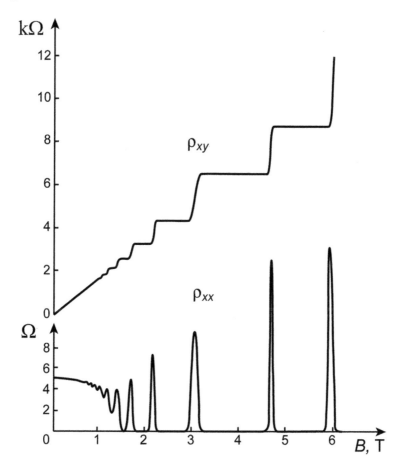

Figure 15.3: Quantization of the Hall resistivity in terms of the temperature and magnetic field.

Each of these plateau is centered on a magnetic field value equal to:

$$B = \frac{hn_e}{ne} = \frac{n_e}{n}\Phi_0 \qquad (15.19)$$

where Φ_0 is quantum flux of the magnetic field.

15.2   TWO-DIMENSIONAL ELECTRONIC SYSTEMS

To write the Schrödinger equation of the system, a gauge is required: the most convenient are the Landau gauge and the symmetric gauge. We will also describe the quantization of the levels of Landau and the effects of the impurities in the material.

15.2.1   Hamiltonian for Non-Interacting Electrons

The quantum problem of a two-dimensional electronic system in the presence of an intense magnetic field can be regarded as one where electron-electron interactions are neglected and the spin will not be considered. This choice is explained by the fact that in the presence of an intense magnetic field the Zeeman effect occurs: for free electrons the energy levels associated with different spin values splits as $\Delta = g\mu_B B$ where μ_B is the Bohr magneton and g the Lande factor. This effect can be overlooked because, in the presence of intense magnetic fields, spin effects are negligible compared to those due to the field itself. The Hamiltonian in the quantum case for free electrons is given by:

$$H = \frac{1}{2m_e}\left[\boldsymbol{p} - e\boldsymbol{A}(\boldsymbol{r})\right]^2 + g\mu_B \boldsymbol{s}\boldsymbol{B} \tag{15.20}$$

where $\boldsymbol{p}$ is the moment operator, $\boldsymbol{A}$ is the vector potential corresponding to the magnetic field, and s is the spin quantum number. The second term regarding the spin moment will be overlooked, as will not be considered the electron-electron interactions. The moment $\boldsymbol{p}$ satisfies the commuting relations $[p_i, r_j] = -i\hbar\delta_{ij}$, where $i, j = x, y, z$ and δ_{ij} is the Kronecker delta. Contrary to what happens in classical mechanics this operator is not directly proportional to the speed operator. In fact, the latter in quantum mechanics derives from the Heisenberg's equations of motion:

$$\boldsymbol{v} = \frac{i}{\hbar}[H, \boldsymbol{r}] = \frac{\boldsymbol{p} - e\boldsymbol{A}}{m_e} \tag{15.21}$$

To overcome this problem, the dynamical moment operator π is introduced, proportional to the speed operator:

$$\pi = m_e \boldsymbol{v} = \boldsymbol{p} - e\boldsymbol{A} \tag{15.22}$$

This operator also follows the following commuting relation:

$$
\begin{aligned}
[\pi_x, \pi_y] &= [p_x - eA_x(r), p_y - eA_y(r)] \\
&= e\left([p_x, A_y] - [p_y, A_x]\right) \\
&= -ie\hbar\left(\frac{\partial A_y}{\partial x} - \frac{\partial A_x}{\partial y}\right) \\
&= -ie\hbar\left(\nabla \times \boldsymbol{A}\right)_z \\
&= -ie\hbar B \\
&= -i\frac{\hbar^2}{l^2}
\end{aligned}
\tag{15.23}
$$

where we have introduced the Larmor radius $l = \sqrt{\hbar/|e|B}$ also called magnetic length: this length is a fundamental scale of the problem and typically falls within a range of values between $(50-100\,\text{Å})$. It is important to note also that the length is independent of the geometry and composition of the material considered. To make an example, considering a magnetic field of $1\,T$, the magnetic length results $\approx 2,5 \times 10^{-8}$m.

Returning to the operator π, even these, as well as the operator p does not commutes with the Hamiltonian.

Let us now consider the effects of magnetic field: the latter assumed uniform, the system must be invariant under translation. Then another operator called, translator operator, is introduced, which has to commute with the Hamiltonian. This is given by:

$$
\boldsymbol{K} = \boldsymbol{p} - e\boldsymbol{A} + e\boldsymbol{B} \times \boldsymbol{r}
\tag{15.24}
$$

called the pseudomoment.

The translator's commuting relations is given by:

$$
[K_x, K_y] = -ie\hbar B = i\frac{\hbar^2}{l^2}
\tag{15.25}
$$

This non-commutativity translational operator is an expression of the fact that a motion in a magnetic field induces a phase in the wave function called Aharonov-Bohm.

15.2.2 Symmetric Gauge

To determine the wave function of the system it is necessary to proceed defining the gauge used. For a system under the effect of a magnetic field, a symmetric gauge is chosen. Hamiltonian does not depend directly on the applied magnetic field B but on the potential vector A, whose choice is not unequivocally determine: different choices of the vector potential generate the same magnetic field but different results for the wave function.

For a real and regular function $\chi(r)$, a gauge transformation is defined as:

$$A'(r) = A(r) + \nabla\chi(r) \tag{15.26}$$

where $A(r)$ and $A'(r)$ are expressions of the same magnetic field. Writing Hamiltonian as:

$$H = \frac{1}{2m_e}[p - eA(r)]^2 + H_{\text{pot}} \tag{15.27}$$

where H_p denotes a generic external potential, it is possible to define the eigenfunction $\phi'(r)$ corresponding to the Hamiltonian eigenvalue E. Then eigenfunctions with the same eigenvalue E of a Hamiltonian in which A is replaced by A' are given by:

$$\phi'(r) = \phi(r) \exp\left(\frac{ie}{\hbar}\chi(r)\right) \tag{15.28}$$

Since $\chi(r)$ is a real function, the choice of a gauge does not affect the probability $|\phi(r)|^2$ of finding the electron in r. This invariance of the observable respect to a freedom in the choice of vector potential is defined as gauge invariance.

15.2.3 Landau Levels

To describe the Hamiltonian of a free electron bound to move in a two-dimensional system we use the dynamical moment π. We can write the Hamiltonian as follows:

$$H = \frac{1}{2m_e}\pi^2 \tag{15.29}$$

We have seen that components of the dynamical moment do not commute between them (15.23). The position operator is then introduced using the coordinates center operators (X, Y). We can define such operator as:

$$r = \left(X + \frac{l^2}{\hbar}\pi_y \, , \, Y - \frac{l^2}{\hbar}\pi_x \right) \tag{15.30}$$

These operators are related to the pseudomoment K:

$$K_x = -eBY \, , \qquad K_y = eBX \tag{15.31}$$

Then, they commute with the dynamical moment π and is equal to $[X, Y] = il^2$.

Returning to the Hamiltonian, we can express it in terms of the components of the dynamical moment such as:

$$H = \frac{1}{2m_e}(\pi_x^2 + \pi_y^2) \tag{15.32}$$

From this equation, since the dynamical moment operator appears in a quadratic form, we can notice that Hamiltonian has the same algebraic structure as that of a harmonic oscillator in a dimension. Let us then introduce non-Hermitian, lowering and raising operators, which satisfy the commuting relationship given by $[a, a^\dagger] = 1$ and are given by:

$$a = \frac{l}{\sqrt{2\hbar}}(\pi_x - i\pi_y) \tag{15.33}$$

$$a^\dagger = \frac{l}{\sqrt{2\hbar}}(\pi_x + i\pi_y) \tag{15.34}$$

where l is the magnetic length. We can rewrite the Hamiltonian as:

$$H = \hbar\omega_c \left(a^\dagger a + \frac{1}{2} \right) \tag{15.35}$$

Hamiltonian eigenstates are the same states as the number operator $N = a^\dagger a$, given $a^\dagger a|n\rangle = n|n\rangle$, where lowering and raising operators act as follows:

$$a|n\rangle = \sqrt{n}|n-1\rangle \tag{15.36}$$
$$a^\dagger|n\rangle = \sqrt{n+1}|n+1\rangle \tag{15.37}$$

where $n \in \mathbb{N}$. The state $|0\rangle$ is defined such that $a|0\rangle = 0$. Energy levels are discretized as for the harmonic oscillator:

$$E_n = \hbar\omega_c \left(n + \frac{1}{2} \right) \tag{15.38}$$

Such discrete levels are the so-called Landau levels and are independent of the gauge used. We also note that in the presence of the magnetic field these levels are equidistant with a energy gap proportional to the field.

15.2.4   Degeneration of Levels and System Eigenstates

At this point we notice a discrepancy: starting with considering the motion of a particle in a plane, having two degrees of freedom, we have come to a result that is that of a harmonic oscillator in a dimension. This is explained by the fact that we have not considered the degeneration of energy levels.

To define the degeneration of the levels we start by considering the symmetric gauge $A = (-By/2, Bx/2, 0)$. This gauge choice implies new advantageous commuting relations for the operators.

Following the choice of the gauge we can express the angular momentum of the system as:

$$L_z = -\frac{\hbar}{2l^2}(X^2 + Y^2) + \frac{l^2}{2\hbar}(\pi_x^2 + \pi_y^2) \tag{15.39}$$

Since $[L_z, H] = 0$, angular momentum eigenvalues can be used to distinguish states related to the same energy level. It is now necessary to introduce new scale operators $b, b^\dagger$ that respect the following commuting relations:

$$[b, b^\dagger] = 1, \qquad [a, b] = [a, b^\dagger] = 0, \qquad [b, H] = [b^\dagger, H] = 0 \tag{15.40}$$

These operators are defined by:

$$b = \frac{1}{\sqrt{2}l}(X + iY) \tag{15.41}$$

$$b^\dagger = \frac{1}{\sqrt{2}l}(X - iY) \tag{15.42}$$

It is also possible in this case to introduce the number operator $b^\dagger b|m\rangle = m|m\rangle$ from which we obtain the new quantum number m, which acts on the analytical form of the wave function, leaving the energy spectrum unchanged.

Once defined such operators we can write the angular momentum as:

$$L_z = \hbar(a^\dagger a - b^\dagger b) \tag{15.43}$$

Therefore, the eigenvalue of the angular momentum will be given by $\hbar(n - m)$. It becomes clear how Landau's levels are degenerate: in a 2-d system the particle has two degrees of freedom associated with the different quantum numbers n, m. But since Hamiltonian is dependent only on the quantum number n, this implies that all states with n fixed and different quantum numbers m have the same energy and are therefore are degenerate.

In the symmetric gauge, each level of Landau contains degenerate orbitals given by the quantum number m. What remains equal for each level is degeneration per unit of surface given by:

$$n_B = \frac{1}{2\pi l^2} \tag{15.44}$$

In order to be able to write the waveforms of the system eigenstates, we begin with the concept that in the symmetric gauge the state of an electron is described by a ket $|n, m\rangle$ where $n, m \leq 0$ are the main quantum numbers of the system. This eigenstates are simultaneously eigenstates of the operators $a, a^\dagger, b$ and $b^\dagger$, that is $a^\dagger a|n, m\rangle = n|n, m\rangle$; $b^\dagger b|n, m\rangle = m|n, m\rangle$. In order to obtain the wave functions expressed as a function of the coordinates of the system we must first express the scale operators according to the latter as:

$$a = -\frac{1}{\sqrt{2}l}\left[\frac{i}{2}(x - iy) + il^2\left(\frac{\partial}{\partial x} - i\frac{\partial}{\partial y}\right)\right] \tag{15.45}$$

$$b = \frac{1}{\sqrt{2}l}\left[\frac{i}{2}(x + iy) + l^2\left(\frac{\partial}{\partial x} + i\frac{\partial}{\partial y}\right)\right] \tag{15.46}$$

Assuming the magnetic field B in the opposite direction, it is possible to introduce the complex coordinates $z = (x - iy)/l$ and $z^* = (x + iy)/l$. We can now rewrite the scale operators according to these coordinates such that:

$$a = -i\sqrt{2}\exp\left(-\frac{|z|^2}{4}\right)\frac{\partial}{\partial z^*}\exp\left(\frac{|z|^2}{4}\right) \tag{15.47}$$

$$a^\dagger = \frac{i}{\sqrt{2}}\exp\left(-\frac{|z|^2}{4}\right)\left[z^* - 2\frac{\partial}{\partial z}\right]\exp\left(\frac{|z|^2}{4}\right) \tag{15.48}$$

And for other operators:

$$b = \sqrt{2} \exp\left(-\frac{|z|^2}{4}\right) \frac{\partial}{\partial z} \exp\left(\frac{|z|^2}{4}\right) \tag{15.49}$$

$$b^\dagger = \frac{1}{\sqrt{2}} \exp\left(-\frac{|z|^2}{4}\right) \left[z - 2\frac{\partial}{\partial z^*}\right] \exp\left(\frac{|z|^2}{4}\right) \tag{15.50}$$

The wave functions of the fundamental state $|0,0\rangle$ is obtained from the conditions $a|0,0\rangle = b|0,0\rangle = 0$. Since $\langle r|0,0\rangle \equiv \psi_{0,0}(r)$, then we can write:

$$\psi_{0,0}(r) = \frac{1}{\sqrt{2\pi}l} \exp\left(-\frac{|z|^2}{4}\right) = \frac{1}{\sqrt{2\pi}l} \exp\left(-\frac{r^2}{4l^2}\right) \tag{15.51}$$

The wave functions of the successive states are obtained starting from the fundamental state and applying the raising operators:

$$\psi_{n,m}(r) = \frac{a^{\dagger n} b^{\dagger m}}{\sqrt{n!}\sqrt{m!}} \psi_{0,0}(r) \tag{15.52}$$

The wave function corresponding to the lowest energy level, where $n = 0$:

$$\psi_{0,m}(r) = \frac{1}{l\sqrt{2\pi 2^m m!}} z^m \exp\left(-\frac{|z|^2}{4}\right)$$
$$= \frac{1}{l\sqrt{2\pi 2^m m!}} \left(\frac{x - iy}{l}\right)^m \exp\left(-\frac{r^2}{4l^2}\right) \tag{15.53}$$

This wave function represents a circularly localized electron so that the maximum probability density, given by $|\psi_{0,m}(r)|^2$, lies along a radius of circumference $l\sqrt{2m}$.

The expectation value of r^2 is given by $\langle 0,m|r^2|0,m\rangle = 2(m+1)l^2$. The state $|0,m\rangle$ should be considered as a linear combination of several cyclotronic orbits of radius l having a center in a circular radius orbit $l\sqrt{2m}$.

15.2.5 Aharonov-Bohm Quantization and Phase Condition

The state corresponding to the lowest energy level is circularly located. What is important to observe is how the wave function phase changes along this circumference. The phase changes by $-2\pi m$: this is the quantization condition of the angular momentum. The effect can be seen as a result of the gauge invariance for the vector potential A: this implies that a charged particle that travels along a closed curve in the presence of a magnetic field acquires a phase:

$$\varphi = \frac{e}{\hbar} \oint A \cdot ds = \frac{e}{\hbar} \int B_n \cdot dS = \frac{e}{\hbar} \Phi \qquad (15.54)$$

From the Stokes theorem and the relation $\nabla \times A = B$. We can therefore see how the phase difference $\Delta\varphi$ between any pair of trajectories having the same end point is determined by the magnetic flux Φ that crosses the surface enclosed by the trajectory of the particles.

That phase, known as Aharonov-Bohm phase, is therefore dependent only on the total flow enclosed by the trajectory and not of the magnetic field passing through it. Since this phase is defined as the phase change due to the angular momentum, we can say that a state with a finite angular momentum must be associated with to corresponding magnetic flux.

15.2.6 Motion in an External Electric Field and Landau Gauge

Let's start by considering a two-dimensional electronic system subjected to the action of an external electric field E along the x axis. In this case, the angular momentum is not conserved and the symmetric gauge is no longer a convenient: then we introduce the Landau gauge:

$$A = (0, B_x, 0) \qquad (15.55)$$

We can then write the Hamiltonian as:

$$H = \frac{1}{2m_e} \left[p_x^2 + (p_y - eB_x)^2 \right] - eE_x \qquad (15.56)$$

This gauge holds the moment p along the axis of the ordinates, which allows us to write the wave function as follows:

$$\psi(r) = \frac{1}{\sqrt{L_y}} \exp(ik_y y)\psi(x) \qquad (15.57)$$

where L_y si the angular momentum along the y direction.

From the previous equation we can then write the Schrödinger equation,

$$\left\{ \frac{1}{2m_e} \left[p_x^2 + (\hbar k_y - eB_x)^2 \right] - eE_x \right\} \psi(x) = \Sigma \psi(x) \tag{15.58}$$

Here Σ indicates the eigenvalue associated to the eigenstate $\psi(x)$. For convenience, we introduce the center of potential,

$$X = -k_y l^2 + \frac{e m_e l^4 E}{\hbar^2} \tag{15.59}$$

So that, we can rewrite (15.58) as:

$$\left[\frac{1}{2m_e} p_x^2 + \frac{m_e \omega_c^2}{2} (x - X)^2 \right] \psi(x) = \left[\Sigma + eEX - \frac{m_e}{2} \left(\frac{E}{B} \right)^2 \right] \psi(x) \tag{15.60}$$

Schrödinger's equation is similar to the harmonic oscillator equation, whose potential is centered in X. It is possible to write the eigenfunction of the system as follows:

$$\psi(x) = \left(\frac{1}{\pi} \right)^{\frac{1}{4}} \left(\frac{1}{2^n n! \, l} \right)^{\frac{1}{2}} \exp\left(-\frac{(x - X)^2}{2l^2} \right) H_n \left(\frac{x - X}{l} \right) \tag{15.61}$$

where H_n are the Hermite polynomials, given by

$$H_n(x) = (-1)^n \exp(x^2) \frac{d^n}{dx^n} \exp(-x^2) \tag{15.62}$$

The energy eigenvalue is given by:

$$\Sigma = \hbar \omega_c \left(n + \frac{1}{2} \right) - eEX + \frac{m_e}{2} \left(\frac{E}{B} \right)^2 \tag{15.63}$$

The first term corresponds to the energy's eigenvalue of a harmonic oscillator, the second term is the energy potential due to the presence of the field and the last term is the kinetic energy due to the translational motion. The expectation value of the speed operator along the two main directions is given by:

$$\langle\psi|v_x|\psi\rangle = \langle\psi|\left(\frac{p_x}{m_e}\right)|\psi\rangle = 0 \tag{15.64}$$

$$\langle\psi|v_y|\psi\rangle = \langle\psi|\left[\frac{1}{m_e}(p_y - eBx)\right]|\psi\rangle = \langle\psi|\left[\frac{eB}{m_e}(X - x) - \frac{E}{B}\right]|\psi\rangle$$

$$= -\frac{E}{B} \tag{15.65}$$

The electronic motion takes place perpendicular to the electric field at a speed E/B just as in classical mechanics. The wave function extends along this equatorial line and its width is in the order of $l\sqrt{2n+1}$.

15.2.7 Effects of Impurities

We have considered so far the case of a pure and therefore ideal sample. In the QHE study, however, we encounter problems with the impurities in the sample materials.

Impurities, being essentially atoms, molecules or ions of different elements, generate interactions between the electrons and the crystalline lattice. This phenomenon is obviously due to the potential difference of impurities which must then be added to the Hamiltonian.

To study the effects of impurities, it is generally considered the mean free path λ and the relaxation time τ, defined as the distance and the average time between a collision and the next.

We will assume that the collisions between the electrons and the lattice are perfectly elastic. The first correction refers to the eigenvalues of the energy considered in the preceding paragraphs: the effects of impurities can be considered as summing the complexity of the harmonic oscillator to a complex term equal to:

$$E_n = \hbar\omega_c\left(n + \frac{1}{2}\right) + i\frac{\hbar}{\tau} \tag{15.66}$$

This way each Landau level is subdivided into a set of energy levels characterized by a Lorentzian density.

However, this is an approximation. The Lorentzian distribution is useful for sufficiently small values of $\omega_c\tau$ but fails for high intensity magnetic fields.

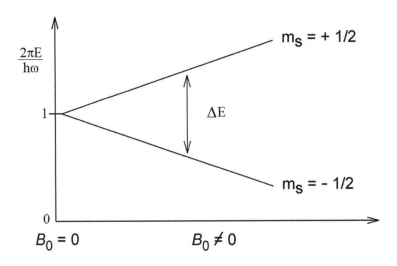

Figure 15.4: Splitting of the levels due to a magnetic field.

We observed how the motion of electrons follows a cyclotron trajectory. But if impurities are present in small quantities and are sufficiently spaced in comparison to the radius of the trajectory, then occurs that a certain number of electrons will continue in its motion without energy changes as opposed to those having the potential of impurities. The variation of the eigenstates energy will be equal to $\pm n\hbar\omega_c$ depending on whether the potential is repulsive or attractive.

For some electrons there is no motion change, except for the fact that the cyclotron beam is replaced by the magnetic length l.

Other effects include electric current. At low temperatures, generally used for QHE, electrons populate all states of energy lower than Fermi's energy. Due to impurities, however, the electrons can be in a localized state or in an extended state. At low temperatures only extended states particípate in electrical conduction. These states are organized in the material with a bend structure as shown in the figure:

What is to be noted is that the presence of impurities reduces the number of extended states by turning into localized states, so the number of states in which the conduction occurs is reduced.

The density of extended states and the plateaus present in Hall's resistance are closely related to the magnetic field. We know that in the absence

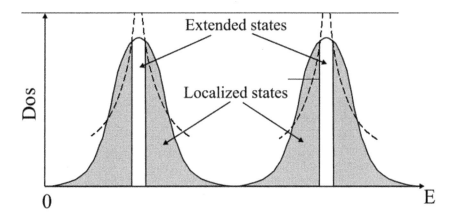

Figure 15.5: Distribution of localized states as a function of energy.

of a magnetic field, the density of states is a constant function of energy. The presence of the field generates a separation of the levels given by the cyclotron energy: in this way there are created forbidden level between Landau levels.

Increasing the intensity of the field increases the separation between the levels, whose energy value approaches to Fermi's energy, thus increasing the intensity of the field results in a decrease in the number of fully extended states beyond Fermi level. Once this value has passed, the states no longer participate in the conduction. When Fermi's energy level falls between two Landau levels, electrons cannot jump to the next level: in these conditions, longitudinal resistance becomes null and cross resistance remains constant forming the plateau.

So far it is clear that the Hall resistance in the classical case, is simply given by B/en_e while the number of states participating in the conduction at a Landau level is eB/h. So if there are d Landau levels completely filled below the Fermi energy, which participate in the conduction, we will have a transverse resistance that will be exactly equal to $h/e^2 d$

Contrary to what one would expect, the presence of impurities is indispensable for observing the whole quantization and therefore the plateaus. In fact it is noted that by increasing, within reasonable limits, the amount of disorder, the plateaus become more evident. If we consider a sample without impurities, they disappear completely.

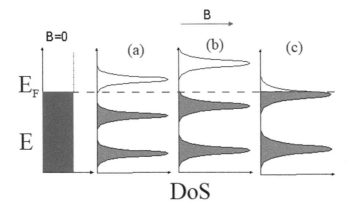

Figure 15.6: Transverse resistance remains constant while Fermi energy is between two levels, cases (a) and (b), and changes when Fermi energy falls within a Level, case (c). In this case we are assuming that the Fermi level is fixed and that only the intensity of the field changes.

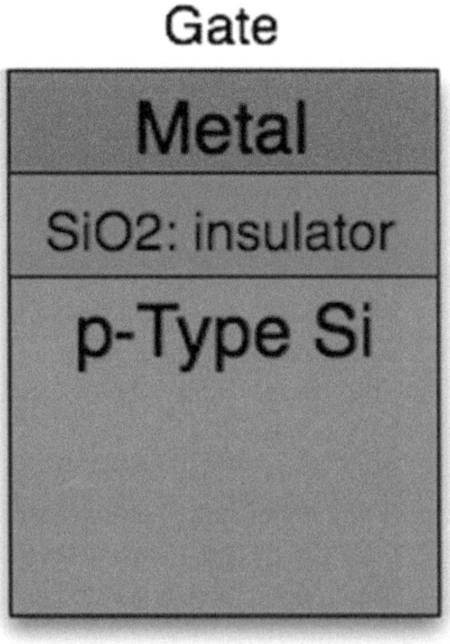

Gate

Metal
SiO2: insulator
p-Type Si

Figure 15.7: Schematic representation of a Si MOS system.

15.3   THE QUANTUM HALL EFFECT—INTEGER QUANTUM HALL EFFECT

We will start with the description of the experimental device used in the experiment where the integer quantum Hall effect (IQHE) was discovered.

We will then address the issue of temperature effects on conductivity and resistivity and ultimately analyze the so-called edge effects.

The IQHE was discovered in a Si MOSFET system. In this systems electrons are trapped in the "inversion band" between the insulating material layer and the semiconductor, whose width is about 30 Å.

GaAs-GaAlAs heterojunctions are more recent two-dimensional electronic systems that and were used in the discovery of the fractional quantum Hall effect: the FQHE. These systems have the electronic density and size of the sample as a common factor.

The first two-dimensional electronic system was built in the 60's with a silicon field MOS transistor. The acronym MOS comes from the three components of the structure: metal, oxide and semiconductor.

The system is designed by combining S_i, a three-dimensional semiconductor, an SiO_2 silicon insulator, and a metal. The important role is played by the semiconductor layer. Silicon is a tetravalent element, i.e., capable of accepting or donning four electrons, and has a crystalline structure very similar to diamond thus formed by covalent bonds.

We know that in a perfect semiconductor the valence band is completely full, while the conduction band is left empty. For silicon the energy of the forbidden band, the gap between the two bands is of about 1.17 eV, at room temperature the gap is much wider than the thermal energy $k_B T$ where $k_B = 1.38065 \times 10^{-23} J/K$ is the Boltzmann constant, so the material acts as an insulator. By working at lower temperatures and by doping silicon, it acquires the properties of a conductor. We can dope the material with donor or doped type n, pentavalent impurities, such as phosphorus P or arsenic As, or with pectin or p-type doped, trivalent impurities, such as boron B or indium In.

Consider an acceptor: the trivalent atom needs an electron, this is taken from the valence band VB and the atom is charged negatively. The captured electron leaves in the VB a "hole," called gap, that interacts with the negative charged accepter, thus creating a bound state. The expression of the bonding energy between the electron and the accepting atom can be expressed as in the case of the hydrogen atom, with the mass correction in this case given by the effective mass and where Coulomb's energy is expressed by the dielectric constant of the semiconductor. The binding energy has the following form:

$$E_a = \frac{e^4 m^*}{2(4\pi\epsilon\hbar)^2} \qquad (15.67)$$

where m^* is the effective mass of the electron in the Si and $\epsilon = 11.7\epsilon_0$ is the dielectric constant of the material. The lower bound state of energy lies above the E_a, where we have indicated the binding energy in the case of an acceptor.

A voltage V_G is applied between the metal and the semiconductor. Suppose that the temperature coincides with absolute zero: the semiconductor Fermi energy is doped p, and lies just above the lower state. We can distinguish three cases. If $V_G = 0$ then the Fermi energy coincides with the semiconductor layer. Other possibility is if $V_G > 0$, and the metal potential is higher that the semiconductor, then the system behaves like a capacitor. In fact between the metal and the insulator, which is the SiO_2 oxide, a positive charge is created, while in the interface between the latter and the semiconductor there is a negative charge. If V_G is weak then the negative charge is given by the accepter near the interface that have gained a negative charge. These are confined to an area called "depletion layer," due to the electric field perpendicular to the layer, generated by V_G, the energy levels bend toward the interface. On the other hand if $V_G \gg 0$ then the change in energy levels increases, this causes a shift of the conduction band. This in fact overcomes the Fermi's level. Then, near the interface, electrons go to those levels of conduction band CB, which are below Fermi's energy. They are confined to a two-dimensional layer along the interface. This area is called "inversion zones," and is essential for observing IQHE.

The thickness of the depletion zones is of a few microns, while the inversion layer is in the order of 10 nanometers. For this reason, we can consider the uniform electric field and the potential as linear, while in the insulating layer the potential is infinite, that is why the electronic motion is confined to a two-dimensional layer.

By changing the applied voltage V_G, it is possible to change the electronic density of the system. Since the conductivity is directly proportional to this density, the system can be used as a transistor. In these systems the density used is about $10^{15} - 10^{17} m^{-2}$, at lower density the effect of impurities becomes too strong. The disadvantage in these systems is given

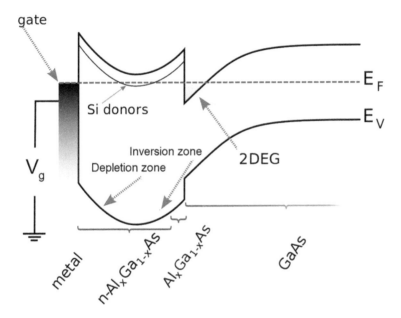

Figure 15.8: Representation of energy levels in a MOSFET sample.

by the depletion zone, which being formed by negative charge receptors, behaves as a source of dispersion potential.

15.3.1 The GaAs-AlGaAs Heterojunctions

This system exploits the properties of gallium Ga and arsenic As, trivalent and aluminum Al, pentavalent. The compound GaAs is on average tetravalent and forms a crystalline structure such as the diamond, in which each gallium atom is bound to four arsenic atoms and vice versa. This is also an insulator, with the Fermi energy level between the valence band and the conduction band. The $Al_xGa_{1-x}As_x$ compound, in which a portion of the gallium is replaced by aluminum, has a similar crystalline structure but a wider energy gap. It is therefore even more insulating. This creates the heterogeneity given by the two sub-layers with different gaps.

Suppose we also work in this case at an absolute zero temperature and introduce it into the donor system. Is to be noted that the Fermi level of substrate $Al_xGa_{1-x}As_x$ lies just above the first bonded donor state which is higher than the bottom of the conducting band of the substrate GaAs.

In this system donor-bound electrons move to the substrate GaAs conduction band. The different charges of the two substrates causes an electric polarization, in fact, the GaAs substrate is charged negatively and $Al_xGa_{1-x}As_x$ positively. The process stops when a certain amount of electrons has moved. One of the advantages of this structure with respect to the Si MOS system is the weaker dispersion potential. Inserting pentavalent elements far from the interface, the effect can be further reduced. This allows to have a high mobility sample. By applying an electrical field perpendicular to the structure, it is possible to control the electronic density, whose utility is similar to a Si MOSFET system.

The results obtained by von Klitzing, Dorda and Pepper in 1980 soon sparked the interest of the international metrology community [139]. In fact, they opened up the possibility to consider the entire quantum Hall effect as a standard measure of resistance measurement, since the latter is dependent only on fundamental constants and is also determined with a high precision.

The official value of this standard resistance, obtained for $i = 1$, is:

$$R_H = \frac{h}{e^2} = 25\,812.807\ \Omega \tag{15.68}$$

This is also called von Klitzing constant.

The finite dimensions of electronic devices have consequences both on the electronic motion and the current that circulates in the sample.

To describe the motion of electrons to the edge of the sample, consider the edge itself as a confinement potential: obviously it affects the state of the electrons in the system. Suppose the system has a rectangular shape in which the largest side lies along the y axis; moreover, the potential for confinement is:

$$V(x) \begin{cases} = 0\,, & \text{if} \quad x < x_0 \\ > 0\,, & \text{if} \quad x > x_0 \end{cases} \tag{15.69}$$

In this situation it is useful to describe the system with the Landau gauge $\boldsymbol{A} = (0, B_x, 0)$. The angular momentum along the y direction is therefore conserved and the wave function is a flat wave. If $V(x)$ varies slowly and continuously, we can use the wave function described in the Eq. (15.61). The electric field is given by the gradient of the potential. We can expand $V(x)$ in Taylor series and since this varies slowly, the second derivative

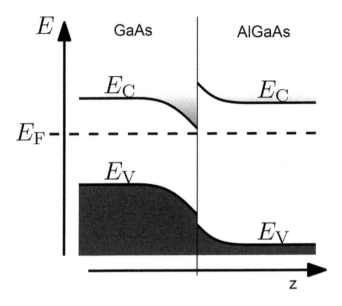

Figure 15.9: Heterojunction representation of the band energy for AlGaAs–GaAs.

and the higher order ones can be neglected within the limit $\delta_x = l\sqrt{2n+1}$. The electron energy is then given by Eq. (15.63).

$V(x)$ could vary slightly faster, then, we can no longer neglect Taylor's terms of higher order. If the function $V(x)$ is quadratic then, we may just approximate the series to second order: in this case, the system is reduced to that of a harmonic oscillator and we are able to solve the Schrödinger equation analytically. The case when $V(x)$ varies very quickly, is the more complicated, it is no longer possible to solve the system analytically. Numerical methods are used. The results of these methods show that the energy of an electronic state increases as it approaches the edge of the sample and in that state, an electron does moves parallel to the edge itself. The motion of the electron can be understood in classically if considered to have reflection at the edge of the cyclotron orbit: the orbit is "broken" at contact with the edge potential barrier, creating an collision that we consider elastic.

15.3.2   Distribution of Current at the Edges of the System

It is important to note that the edge current flows in absence of an external potential too, but in this case the currents on the opposing sides are canceled. In laboratory systems, it is noted that the current passes through the whole sample but not uniformly. On the other hand, to explain the presence of plateau in Hall's resistance, we have to assume the presence of localized states throughout the sample. Fermi level must lie between the Landau levels: more precisely the energy of Fermi must fall in a range other than the energy of extended states. If these conditions do not occur, we cannot ignore the dispersion to the opposite edges of the sample.

However, for an ideal system without impurities, the density of edge states is much lower than that the rest of the system, and the conditions above are verified in the case of the integers of $\hbar/e^2$. With a change in coordinate center from X to l and in the case of a high potential barrier, the energy of edge states varies in the order of $\hbar\omega_c$. Therefore, the ratio between the number of edge states and those of the remaining system is l/L, where L is the sample size. The values for which the Fermi level lies in a range other than the extended states is:

$$\frac{N-l}{L} < i < N \tag{15.70}$$

where N is the number of edge states, corresponding to the number of Landau levels below Fermi energy.

There is, however, a limit to which dissipation grows in various orders of magnitude: this limit is called critical current I_c and has depends of system temperature.

15.3.3   Temperature Effects

Temperature has negligible effects on electronic motion and therefore on the energy levels: this results in variations in the longitudinal resistivity ρ_{xx}, which in the IQHE must tend to zero. Up to 1 K these effects are negligible and the theory seen so far suits the description of the system. The longitudinal resistivity is so reduced that it makes the measure difficult. Exceeding this value however, we begin to observe variations in these magnitudes due to the thermal instability of the system. In fact, the thermal energy kT becomes significant and the longitudinal resistivity has a dependence on the temperature that goes like:

$$\rho_{xx} \propto \exp\left(-\frac{\Delta E}{2kT}\right) \tag{15.71}$$

where ΔE is the energy separation between Landau levels.

For a MOSFET system the resistance is minimal for temperatures below 2 K. Above these values the gap ΔE is equal to 85% of the cyclotronic energy $\hbar\omega_c$ in the case of $i = 2$ and 96% for $i = 4$. For heterojunctions GaAs–AlGaAs the gap is even higher than the cyclotronic energy of 20%. For separating levels due to the magnetic field, we would expect lower values.

The temperature also has effects on the Hall's resistance. Around 1983, it was observed that by measuring ρ_{xy} depending on the values of magnetic field or the applied voltage, Hall's resistivity had a temperature dependence. It was precisely observed that when the temperature increases, ρ_{xy} decreases. The variation $\Delta\rho_{xy} = \rho_{xy}(T) - \rho_{xy}(0)$ actually showed this dependency:

$$\Delta\rho_{xy} = -\alpha\rho_{xx} \tag{15.72}$$

where the parameter α depends on the used device. This dependence is therefore linear.

However, it is important to note that the mechanism of this equation has not been explained. The temperature also has effects on the current circulating in the system. When the applied voltage increases suddenly, it goes to what is called "breakdown" of the IQHE: there is a current limit, the critical current I_c, which causes this sudden voltage variation.

This value is dependent on the system width and the critical current density of 0.5–2.0 A/m. Instead of defining a critical current, it is possible to express everything in terms of the electric field: the two magnitudes are in fact linked by the conductivity of the system.

BIBLIOGRAPHY

1. Kenneth Denbigh. (1981). *The Principles of Chemical Equilibrium with Applications in Chemistry and Chemical Engineering*. Cambridge University Press, Cambridge, Fourth edition.

2. Cohen-Tannoudji, C. (1977). *Quantum Mechanics*, Vols. 1 and 2, Wiley, Paris.

3. Winkler, R. (2003). *Spin-Orbit Coupling Effects in Two Dimensional Electron and Hole Systems*. Springer.

4. Keldysh, L. V. (1964, 1965). Diagram technique for nonequilibrium processes. *Zh. Eksp. Teor. Fiz. 47*, 1515; *JETP Lett. 20*, 1018.

5. Guinea, F., Huertas-Hernando, D., & Brataas, A. (2006). Spin-orbit coupling in curved graphene, fullerenes, nanotubes, and nanotube caps, *Phys. Rev. B, 74*, 155426.

6. Katsnelson, I. (2012). *Graphene, Carbon in Two Dimensions*, Cambridge University Press.

7. Dresselhaus, G. (1955). Spin-orbit coupling effects in zinc blende structures. *Phys. Rev., 100*, 580.

8. Rashba, E. I., & Bychkov, Y. A. (1984). Oscillatory effects and the magnetic susceptibility of carriers in inversion layers. *Sov. Phys. Solid State, 17*, 6039.

9. Imry, Y., Büttiker, M., & Landauer, R. (1983). Josephson behavior in small normal one-dimensional rings . *Phys. Lett. A*, 96.

10. López, A., Bolívar, N., Medina, E., & Berche, B. (2010). A perfect spin filtering device through MachZehnder interferometry in a GaAs/AlGaAs electron gas. *Journal of Physics Condensed Matter, 22*, 115303.

11. Medina, E., Ellner, M., Bolívar, N., & Berche, B. (2014). Charge- and spin-polarized currents in mesoscopic rings with Rashba spin-orbit interactions coupled to an electron reservoir. *Physical Review B*.

12. Tokatly, I. V. (2008). Equilibrium Spin Currents: Non-Abelian Gauge Invariance and Color Diamagnetism in Condensed Matter. *Phys. Rev. Lett., 101*, 106601.

13. Medina, Berche, B., & López, A. (2012). Spin superfluidity and spin-orbit gauge symmetry fixing. *Europhysics Lett., 97*, 67007.

14. López, A., Berche, B., Bolívar, N., & Medina, E. (2009). Gauge field theory approach to spin transport in a 2D electron gas. *Condensed Matter Physics, 12*, 707.

15. Bolívar, N., López, A., Medina, E., & Berche, B. (2009), Comment on "Equilibrum spin currents: Non-Abelian Gauge Invariance and Color Diamagnetism in Condensed Matter". *arXiv:cond-mat/0902.4635.*, Preprint.

16. Castro Neto A. H. et al. (2009). The electronic properties of graphene. *Rev. Mod. Phys., 81*, 109.

17. Guinea, González, J., & Herrero, J.(2009). Propagating, evanescent, and localized states in carbon nanotube – graphene junctions *Phys. Rev. B, 79*, 165434.

18. Kane, C. L., & Mele, E. J. (1997). Size, shape, and low energy electronic structure of carbon nanotubes. *Phys. Rev. Lett., 78*, 1932.

19. Akhmerov, A. R., & Beenakker, C. W. J. (2008). Boundary conditions for Dirac fermions on a terminated honeycomb lattice. *Phys. Rev. B, 77*, 085423.

20. Medina, E., Bolívar, N., & Berche, B. (2014), Persistent charge and spin currents in the long-wavelength regime for graphene rings. *Phys. Rev. B, 89*, 125413.

21. Foldy, L., & Wouthuysen, S. A. (1950). On the Dirac Theory of Spin 1/2 Particles and Its Non-Relativistic Limit. *Physical Review, 78*, 29.

22. Mohanty, P. (1999). Persistent current in normal metals. *Ann. Phys., 8*, 549.

23. Chandrasekhar, V. et al. (1991). Magnetic response of a single, isolated gold loop. *Phys. Rev. Lett., 67*, 3578.

24. Aharonov, Y., & Bohm, D. (1959). Significance of electromagnetic potentials in the quantum theory. *Phys. Rev., 115*, 485.

25. Ehrenberg, W., & Siday, R., (1949). The Refractive Index in Electron Optics and the Principles of Dynamics. *Proceedings of the Physical Society. Series B, 62*, 8.

26. Datta, S. & Das, B. (1990). Electronic analog of the electrooptic modulator. *Appl. Phys. Lett., 56*, 665.

27. Burkov, A.A. et al. (2004). Theory of spin-charge-coupled transport in a two-dimensional electron gas with Rashba spin-orbit interactions. *Phys. Rev. B, 70*, 155308.

28. Zhang, S., & Yang, Z. (2005). Intrinsic spin and orbital angular momentum Hall effect. *Phys. Rev. Lett., 94*, 066602.

29. Xiao, D., Shi, J., Zhang, P., & Niu, Q. (2006). Proper definition of spin current in spin-orbit coupled systems. *Phys. Rev. Lett., 96*, 076604.

30. Ryder, L. H. (1985). *Quantum Field Theory*, Cambridge University Press.

31. Rebei, A,.& Heinonen, O., (2006). Spin currents in the Rashba model in the presence of nonuniform fields. *Phys. Rev. B, 73*, 153306.

32. Li, Y., Jin, P. Q., & Zhang, F. C. (2006). $SU(2) \times U(1)$ unified theory for charge, orbit and spin currents. *Phys. A: Math. Gen., 39, number 22*, 7115.

33. Goldhaber, A. S. (1989). Comment on "Topological Quantum Effects for Neutral Particles." *Phys. Rev. Lett., 62*, 482.

34. Frohlich, J., & Studer, U. M. (1993). Gauge invariance and current algebra in nonrelativistic many-body theory. *Rev. Mod. Phys., 65*, 733.

35. Santiago, D. I., Leurs, B. W. A., Nazario, Z., & Zaanen, J. (2008). Corrigendum to "Non-Abelian hydrodynamics and the flow of spin in spin–orbit coupled substances." *Ann. Phys., 323*, 907.

36. Rashba, E. I. (1960). Properties of semiconductors with an extremum loop .1. Cyclotron and combinational resonance in a magnetic field perpendicular to the plane of the loop. *Sov. Phys. Solid State, 2*, 1109.

37. Nitta, J., & Koga, T. (2003). Rashba spin–orbit interaction and its applications to spin-interference effect and spin-filter device. *J. Supercond., 16*, 689.

38. Fabian, J., Zutić, I., & Das Sarma, S. (2004). Spintronics: Fundamentals and applications,. *Rev. Mod. Phys., 76*, 323.

39. Usaj, G., & Balseiro, C. A. (2005). Spin accumulation and equilibrium currents at the edge of 2DEGs with spin-orbit coupling. *Europhys. Lett., 72*, 631.

40. Peskin, M. E. & Schröder, D. V. (1995). *Quantum Field Theory*.

41. Weinberg, S. (1996). *The Quantum Theory of Fields*, Vol. II. Cambridge University Press.

42. Rashba, E. I. (2004). Sum rules for spin Hall conductivity cancellation. *Phys. Rev. B, 70*, 201309.

43. T. Koga, J. Nitta, & M. van Veenhuizen. (2004). Ballistic spin interferometer using the Rashba effect. *Phys. Rev. B, 70*, 161302.

44. Chen, S. H. & Chang, C. R. (2008). Non-Abelian spin-orbit gauge: Persistent spin helix and quantum square r. *Phys. Rev. B., 77*, 045324.

45. Ting, D. Z. Y. & Cartoixa X. (2003). Bulk inversion asymmetry enhancement of polarization efficiency in nonmagnetic resonant-tunneling spin filters. *Phys. Rev. B., 68*, 235320.

46. Miller J. B. et al. (2003). Gate-Controlled Spin-Orbit Quantum Interference Effects in Lateral Transpor. *Phys. Rev. Lett., 90*, 076807.

47. Studer, M., et al. (2009). Gate-Controlled Spin-Orbit Interaction in a Parabolic GaAs/AlGaAs Quantum Well. *Phys. Rev. Lett., 103*, 027201.

48. Liu, R. C., Oliver, W. D., Kim, J., & Yamamoto, Y. (1999). Hanbury Brown and Twiss-Type Experiment with Electrons. *Science, 284*, 299.

49. Aranzana, M., Feve, G., Oliver, W. D., & Yamamoto, Y. (2002). Rashba effect within the coherent scattering formalism. *Phys. Rev. B, 66*, 155328.

50. Shirasaki, R., Hatano, N., & Nakamura, H. (2007). Non-Abelian gauge field theory of the spin-orbit interaction and a perfect spin filter. *Phys. Rev. A, 75*, 032107.

51. Aharonov, Y., & Casher, A. (1984). Topological quantum effects for neutral particles. *Phys. Rev. Lett., 53*, 319.

52. Chatelain, C., Berche, B., & Medina, E. (2010). Mesoscopic rings with spin-orbit interactions. *Eur. J. Phys., 31, number 5*, 1267.

53. Imry, Y., Büttiker, M., & Azbel, M. Ya., (1984). Quantum oscillations in one-dimensional normal-metal rings . *Phys. Rev. A, 30*, 1982.

54. Brey, L., & Fertig, H. (2006). Electronic states of graphene nanoribbons studied with the Dirac equation. *Phys. Rev. B, 73*, 235411.

55. Min, H., et al. (2006). Intrinsic and Rashba spin-orbit interactions in graphene sheets. *Phys. Rev. B, 74,* 165310.

56. Morpurgo, A. F., Meijer, F. E., & Klapwijk, T. M. (2002). One-dimensional ring in the presence of Rashba spin-orbit interaction: Derivation of the correct Hamiltonian. *Phys. Rev. B, 66,* 033107.

57. Cotaescu, I. I. et al. (2007). Signatures of the Dirac electron in the flux dependence of total persistent currents in isolated AharonovBohm rings. *J. Phys.: Condens. Matter, 19*(24), 242206.

58. Berry, M. V., Bender, C. M., & Mandilara, A. (2002). Generalized PT symmetry and real spectra. *J. Phys. A: Math. Gen., 35,* L467.

59. Mecklenburg, M., & Regan, B. C. (2011). Spin and the Honeycomb Lattice: Lessons from Graphene. *Phys. Rev. Lett., 106,* 116803.

60. Nakanishi, T., Ando, T., & Saito, R. J. (1998). Berry's phase and absence of back scattering in carbon nanotubes. *Phys. Soc. Jap., 67,* 2857.

61. Recher, P., et al. (2007). Aharonov-Bohm effect and broken valley degeneracy in graphene rings. *Phys. Rev. B., 76,* 235404.

62. Feynman, R. P. (1965). *The Character of Physical Laws.* Cox and Wyman Ltd., London.

63. Feynman, R. P., Leighton, R. B., & Sands, M. (1963). *The Feynman Lectures on Physics,* Vol. III, Addison-Wesley, Chapter 1.

64. Itzykson, C., & Zuber, J. B. (1980). *Quantum Field Theory,* McGraw-Hill.

65. Kaku, M. (1993). *Quantum Field Theory: A Modern Introduction,* Oxford University Press.

66. Peskin, M. E., & Schröder, D. V. (1995). *An Introduction in Quantum Field Theory.* Perseus Books, Reading Massachusetts.

67. Polyak, M. (2005). Feynman diagrams for pedestrians and mathematicians, in: *Graphs and Patterns in Mathematics and Theoretical Physics,* Edited by: Mikhail Lyubich, and Leon Takhtajan, Proceedings of Symposia in Pure Mathematics, AMS.

68. Ramond, P. (2001). *Field Theory: A Modern Primer,* 2nd ed., Westview.

69. Ryder, L. (1996). *Quantum Field Theory.* Cambridge University Press.

70. Schwartz, L. (1981). *Cours d'analyse* (French Edition).

71. Ticciati, R. (1999). *Quantum Field Theory for Mathematicians,* Cambridge University Press.

72. Nakanishi K. Wakabayashi, T., Sasaki, K., & Enoki, T. (2010). Electronic states of graphene nanoribbons and analytical solutions. *Sci. Technol. Adv. Mater., 11,* 054504.

73. Hwang D. Sarma, E. H., Shaffique Adam, & Rossi, E. (2011). Electronic transport in two-dimensional graphene. *Rev. Mod. Phys., 83,* 407.

74. Egues J. C., et al. (2005). Shot noise and spin-orbit coherent control of entangled and spin-polarized electrons. *Phys. Rev. B, 72,* 235326.

75. Jaen-Seung & Hyun-Woo, (2009). Curvature-enhanced spin-orbit coupling in a carbon nanotube. *Phys. Rev. B, 80,* 075409.

76. Ralph F. Kuemmeth, D. C., Ilani, S., & McEuen, P. L. (2008). Coupling of spin and orbital motion of electrons in carbon nanotubes. *Nature, 452*, 448.

77. Blezynski-Jayich, A. C. et al. (2009). Persistent currents in normal metal rings. *Science*, 326, 272.

78. Ci, L. et al. (2008). Controlled nanocutting of graphene. *Nano Research, 1*, 116–122.

79. Balakrishnan, J. et al. (2013). Colossal enhancement of spin-orbit coupling in weakly hydrogenated graphene. *Nature Physics, 9*, 284.

80. Baym, G. (1990). *Lectures in Quantum Mechanics*, Westview Press, New York.

81. Davies, J. H. (1998). *The Physics of Low-Dimensional Semiconductors: An Introduction*. Cambridge University Press.

82. Lassnig, R. (1985). K-P theory, effective-mass approach, and spin splitting for two-dimensional electrons in GaAs-GaAlAs heterostructures. *Phys. Rev. B, 31*, 8076–8086.

83. Richard, P. Feynman, Robert, B. Leighton, & Matthew, L. Sands, (1989). *The Feynman Lectures on Physics*. Addison-Wesley.

84. Büttiker, M. (1985). Small normal-metal loop coupled to an electron reservoir. *Phys. Rev. B, 32*, 1846–1849.

85. Robert Eisberg & Robert Resnick, (1985). *Quantum Physics of Atoms, Molecules, Solids, Nuclei, and Particles*. Wiley, 2nd edition.

86. Neil W. Ashcroft & David Mermin, N., (1976). *Solid State Physics*. Brooks Cole, 1st edition.

87. Shun L. Chuang, (1995). *Physics of Optoelectronic Devices: Wiley Series in Pure and Applied Optics*. Wiley-Interscience.

88. Charles Kittel. (1986). *Introduction to Solid State Physics*. John Wiley & Sons, Inc., New York, 6th edition.

89. Kane, E. O. (1957). *Journal of Physics and Chemistry of Solids, 1*(4), 249–261.

90. Bastard, G., (1992). *Wave Mechanics Applied to Semiconductor Heterostructures*. EDP Sciences.

91. Hans-Andreas Engel, Emmanuel I. Rashba, & Bertrand I. Halperin, (2006). Theory of Spin Hall Effects in Semiconductors. *arXiv*, cond-mat.m (May), 37.

92. John D. Jackson, (1998). *Classical Electrodynamics*, Third Edition. Wiley.

93. Jackson, J. D., & Okun, L. B., (2000). Historical roots of gauge invariance. *Reviews of Modern Physics, 73*(3), 34.

94. Sakurai, J. J., (1993). *Modern Quantum Mechanics (Revised Edition)*. Addison Wesley, 1st edition.

95. Sangchul Oh & Chang-Mo Ryu, (1995). Persistent spin currents induced by the Aharonov-Casher effect in mesoscopic rings. *Phys. Rev. B, 51*, 13441–13448.

96. Aharonov, Y., & Bohm, D., (1959). Significance of electromagnetic potentials in the quantum theory. *Phys. Rev., 115*, 485–491.

97. Chambers, R. G., (1960). Shift of an Electron Interference Pattern by Enclosed Magnetic Flux. *Phys. Rev. Lett., 5*, 3–5.

98. Aharonov, Y., & Casher, A. (1984). Topological Quantum Effects for Neutral Particles. *Phys. Rev. Lett., 53*, 319–321.

99. Cimmino, A., Opat, G. I., Klein, A. G., Kaiser, H., Werner, S. A., Arif, M., & Clothier, R., (1989). Observation of the topological Aharonov-Casher phase shift by neutron interferometry.*Phys. Rev. Lett., 63*, 380–383.

100. Ekenberg, U., & Gvozdic, D. M., (2007). Analysis of electric-field-induced spin splitting in wide modulation-doped quantum wells. *Physical Review B, 78*(20), 10.

101. Bertrand Berche, Christophe Chatelain, & Ernesto Medina, (2010). Mesoscopic rings with spin-orbit interactions. *European Journal of Physics, 31*(5), 1267–1286.

102. Molnár, B., Peeters, F. M., & Vasilopoulos, P., (2004). Spin-dependent magnetotransport through a ring due to spin-orbit interaction. *Phys. Rev. B, 69*, 155–335.

103. Yoseph Imry. (1997). Introduction to Mesoscopic Physics. *Mesoscopic Physics and Nanotechnology*. Oxford University Press, New York.

104. Arnol'd, V. I. *Mathematical Methods of Classical Mechanics*. Moscow.

105. Berezin, A. (1966). *The Method of Second Quantization*, Academic Press.

106. Pierre Deligne, Pavel Etingof, Daniel S. Freed, Lisa C. Jeffrey, David Kazhdan, John W. Morgan, David R. Morrison, & Edward Witten, editors. (1999). *Quantum Fields and Strings: A Course for Mathematicians*. Vols. 1 and 2. American Mathematical Society, Providence, RI. Material from the Special Year on Quantum Field Theory held at the Institute for Advanced Study, Princeton, NJ, 1996–1997. (also lecture notes available online).

107. Di Francesco, Ph., Mathieu, P., & Sénéchal, D., (1997). *Conformal Field Theory*. Springer, New York.

108. Dolgachev, I. *Introduction to String Theory*, preprint – Ann Arbor., Lecture notes available online: http://www.math.lsa.umich.edu/idolga/lecturenotes.html.

109. Etingof, P. *Mathematical Ideas and Notions of Quantum Field Theory*. preprint – MIT lecture notes available online: http://math.mit.edu/etingof/.

110. Faddeev, L. D. & Yakubovsky, O. A. (2009). *Lectures in Quantum Mechanics for Students in Mathematics*, Leningradskii universitet, 1980. (in Russian). English translation: Lectures on Quantum Mechanics for Mathematics Students – L. D. Faddeev, Steklov Mathematical Institute, and O. A. Yakubovskii, St. Petersburg University – with an appendix by Leon Takhtajan – AMS.

111. Tolbert, S. H., Herhold, A. B., Johnson, C. S., & Alivisatos, A. P. (1994). Comparison of quantum confinement effects on the electronic absorption-spectra of direct and indirect gap semiconductor, nanocrystals. *Phys. Rev. Lett., 73*, 3266–3269.

112. Tolbert, S. H., & Alivisatos, A. P. (1995). The wurtzite to rock-salt structural transformation in CdSe nanocrystals under high-pressure. *J. Chem. Phys., 102*, 4642–4656.

113. Woggon, U. (1997). *Optical Properties of Semiconductor Quantum Dots*; Springer-Verlag: Berlin, Germany.

114. Thaller, Bernd. (2013). *The Dirac Equation*. Springer Science & Business Media.

115. Landau, L. D., & Lifshitz, E. M. (2007). *Statistical Physics*, Vol. 5. Elsevier, Oxford, Third edition.

116. Bailey, R. E., & Nie, S. M. (2003). Alloyed semiconductor quantum dots: Tuning the optical properties without changing the particle size. *J. Am. Chem. Soc., 125*, 7100–7106.

117. Schlosshauer, Maximilian, (2003). Decoherence, the measurement problem, and interpretations of quantum mechanics. *Rev. Mod. Phys., 76*,1267–1305.

118. Zurek, W. H., (1981). Pointer basis of quantum apparatus: Into what mixture does the wave packet collapse?. *Phys. Rev. D, 24*,1516–1525.

119. Efros, Al. L., & Efros, A. L. (1982). Interband absorption of light in a semiconductor sphere. *Sov. Phys. Semicond. 16*, 772–775 .

120. Brus L., (1986), Electronic wave functions in semiconductor clusters: experiment and theory. *J. Phys. Chem. 90*(12), 2555–2560.

121. Mittelstaedt, P. (1998). Can EPR-correlations be used for the transmission of super-luminal signals?. *Ann. Phys., 7*, 710–715.

122. Hirsch, J. E. (1999). Spin Hall effect, *Phys. Rev. Lett. 83*, 1834.

123. E. Schrödinger, (1935), Die gegenwärtige Situation in der Quantenmechanik I-III, *Die Naturwissenschaften, 23*, 807–812; 823–828; 844–849.

124. Mehra, J. (1987). Niels Bohr's discussions with Albert Einstein, Werner Heisenberg, and Erwin Schrdinger: The origins of the principles of uncertainty and complementarity. *Found Phys. 17*, 461. https://doi.org/10.1007/BF01559698.

125. Feynman, R. P. (1948). Space-time approach to non-relativistic quantum mechanics, *Rev. Mod. Phys. 20*, 367.

126. Gerlach, W., & Stern, O. (1922), Das magnetische Moment des Silberatoms, Zeitschrift fr Physik, Volume 9, Issue 1, pp 353–355.

127. Dirac, P. A. M. (1928). The quantum theory of the electron, *Proceedings of the Royal Society A., 23.*

128. Ruderman, M. A., & Kittel, C. (1954), Indirect exchange coupling of nuclear magnetic moments by conduction electrons. *Phys. Rev., 96*, 99–102.

129. Mott, N. F. (1936). The resistance and thermoelectric properties of the transition metals, *Proc. Royal Soc. 156*, 368.

130. Julliere, M., (1975). Tunneling between ferromagnetic films, *Phys. Lett. A, 54*, 225–226.

131. Sommerfeld, A. (1916), Die Feinstruktur der Wasserstoff- und der Wasserstoff-ähnlichenLinien.Sitzungsberichte der mathematisch-physikalischen Klasse der K.B. Akademie derWissenschaften zu München, pp. 459–500.

132. Zeeman, P., (1896). The influence of a magnetic field on radiation frequency. *Proceedings of the Royal Society of London* (1854–1905). 60, 513.

133. Rabi, II, Zacharias, J. R., Millman, S., & Kusch, P. (1938). A new method of measuring nuclear magnetic moment. *Phys Rev 53,* 318.

134. Frisch, R., & Stern, O. (1933). "ber die magnetische Ablenkung von Wasserstoff-moleklen und das magnetische Moment des Protons. I / Magnetic Deviation of Hydrogen Molecules and the Magnetic Moment of the Proton. I". *Z. Phys. 85*, 4–16.

135. Bloch, F., Hansen, W. W., & Packard, M. (1946). Nuclear induction. *Phys. Rev., 69*, 127.

136. Purcell, E. M., Torrey, H. C., & Pound, R. V. (1946), Resonance absorption by nuclear moments in a solid. *Phys. Rev., 69,* 37–38.

137. Hall, E. H. (1879). On a new action of the magnet on electric currents *Am. J. Math. 2,* 287–292.

138. Wakabayashi, J.-I., Kawaji, S. (1978). Hall effect in silicon MOS inversion layers under strong magnetic fields, *Physical Society of Japan, Journal, vol. 44,* 1839–1849.

139. Klitzing, K. V., Dorda, G., & Pepper, M. (1980). New method for high-accuracy Determination of the fine-structure constant based on quantized Hall resistance, *Phys. Rev. Lett. 45,* 494–497.

INDEX